VISTAS OF
SPECIAL FUNCTIONS

VISTAS OF
SPECIAL FUNCTIONS

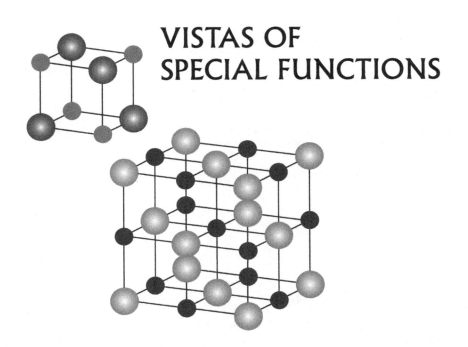

Shigeru Kanemitsu & Haruo Tsukada
Kinki University, Japan

W⊖ **World Scientific**

NEW JERSEY • LONDON • SINGAPORE • BEIJING • SHANGHAI • HONG KONG • TAIPEI • CHENNAI

Published by

World Scientific Publishing Co. Pte. Ltd.

5 Toh Tuck Link, Singapore 596224

USA office: 27 Warren Street, Suite 401-402, Hackensack, NJ 07601

UK office: 57 Shelton Street, Covent Garden, London WC2H 9HE

British Library Cataloguing-in-Publication Data
A catalogue record for this book is available from the British Library.

VISTAS OF SPECIAL FUNCTIONS

For photocopying of material in this volume, please pay a copying fee through the Copyright Clearance Center, Inc., 222 Rosewood Drive, Danvers, MA 01923, USA. In this case permission to photocopy is not required from the publisher.

ISBN-13 978-981-270-774-1
ISBN-10 981-270-774-3

Printed in Singapore.

To Professor Michel Waldschmidt with deep respect

Preface

This book is intended for aspirant readers who are eager to have basic knowledge of special functions in an organic way. We have kept paying attention to make an order in various equivalent statements on special functions. A unique feature is that the reader can gain a grasp of (almost) all existing (and scattered around) formulas in the theory of gamma functions etc. in a clear perspective through the theory of zeta-functions. Thus, this is a book of special functions in terms of the zeta-functions. Reading through this book, the reader can master both fields efficiently. Here a hunter looking for two rabbits gets two.

Here are some descriptions of the contents.

In Chapter 1, we present a unified theory of Bernoulli polynomials with all equivalent conditions properly located. We have revealed that the difference equation (DE) satisfied by the Bernoulli polynomial corresponds to differentiation while the Kubert identity (K) corresponds to integration (the Riemann sum into equal division). This new view point makes the whole theory very lucid.

In Chapter 2 we shall present rather classical and standard theory of the gamma and related functions. Classical as it looks, we shall provide some very unique features of the Euler digamma function from which we may deduce the corresponding properties of the gamma function. Especially, we shall give three proofs of the remarkable formula of Gauss on the values of the digamma function at rational arguments. One is classical and is presented in Chapter 2. Other two proofs are more original given in Chapter 8, one is the limiting case (Theorem 8.2) of the Eisenstein formula in its genuine form (a theorem due to H.-L. Li, L.-P. Ding and M. Hashimoto, describing a basis element in terms of another basis of the space of periodic Dirichlet series), the other is the theorem of M. Hashimoto, S. Kanemitsu and M. Toda about the equivalence between the finite form of the value of the Dirichlet L-function at 1 and the formula of Gauss.

In Chapter 3, we shall present the theory of the Hurwitz zeta-function. The main ingredient is the integral representation for its partial sum. This is to the effect that once we have an integral representation as the one we have, we may immediately draw information for the derivatives, i.e. we have an inheritance of the information. The integral representation for the partial sum is so informative that it contains all information we need (Theorem 3.1). The versatility of this result will be developed in Chapter 5, where through Lerch's formula, we transfer the results on the Hurwitz zeta-function to those on the gamma and related functions. Especially, the asymptotic results established in Chapter 3 will immediately transfer to the Stirling formula and other asymptotic formulas for relatives of the gamma function.

In Chapter 4, we shall present the theory of Bernoulli polynomials through the negative integer values $\zeta(-n, z)$ of the Hurwitz zeta-function. Here we shall establish only three statements, i.e. the Fourier series (H), the difference equation (DE) and the Kubert identity (K) from any of which we may complete the theory following the logical scheme in Chapter 1.

In Chapter 5, first we shall reveal the power of theorems in Chapter 3 to exhibit what the Dufresnoy-Pisot type uniqueness theorem means. Then we shall go on to presenting the first circle (krug p'iervyi) which connects various identities between gamma and trigonometric functions to the functional equations (zeta-symmetry) of the zeta-functions . Thus we shall show that everything comes from the functional equation. A remarkable notice is that such trigonometric identities like the infinite product for the sine function or the partial fraction expansion for the cotangent function are equivalent to the functional equation, thus revealing why Euler succeeded in solving the Basler problem.

In Chapter 6, we shall further pursue this zeta-symmetry in relation to the crystal symmetry through the Epstein zeta-function. We surpass the preceding results by introducing the signs and giving the Chowla-Selberg type formula (based on the Mellin-Barnes integrals) and provide a quick means for computation of the Madelung constants.

In Chapter 7, we shall provide rudiments of the theory of Fourier series and integrals to such an extent that is sufficient for applications and reading through this book, for the sake of the reader who wants to learn it quickly.

Chapter 8 is, so to say, a discrete version of Chapter 7, i.e. the finite Fourier series (transforms). Through this we make clear the orthogonality of characters and other bases of the space of Dirichlet series with periodic coefficients, giving rise to the theorem mentioned above. We can naturally extend our method to develop the similar theory for higher derivatives of the Dirichlet L-function, including Kronecker's limit formula. But because of limitation of time, we cannot go further.

Appendix A gives the very basics of the theory of complex functions. We present mostly results only, and the interested reader should consult a standard book for their proofs. We shall give, however, some details on the use of residue theorem.

Appendix B assembles summation formulas and convergence theorems used in the book. Especially, the Fourier series for the first periodic Bernoulli polynomial is so essential and important, we give two proofs, one depending on ordinary Fourier theory (Chapter 7) and the other on the polylogarithm function of degree 1, where we apply the theorem of Abel and Dirichlet in place of Fourier theory.

As is explained above, Chapters 1 and 4 are parallel, so are Chapters 2 and 5. To understand Chapters 4 and 5, one should read Chapter 3 first. If one finds some difficulties, then one is referred to Appindices A and B. Chapters 7 and 8 can be read independently, but it will be more instructive to read both in parallel. Chapter 6 can be read separately which requires more knowledge of Bessel functions. Because of lack of time, we could not state much about them.

This publication was supported by Kinki University Grant for Publication, No. GK04 in the academic year 2006. The authors are thankful to Kinki University for their generosity of this support. They also would like to thank Ms. Chiew Ying Oi who helped them all through the process with her efficient editorial skills. And toward the end of the process Ms. Zhang Ji supported us and we would like to express our heartily thanks to her.

The authors would like to express their hearty thanks to their close friend Professor Y. Tanigawa for his constant support, encouragement, and stimulating discussions. The first author would like to thank his close friend Professor Heng Huat Chan for his enlightening remark on the equivalent statements to the functional equation, thanks to which he got motivated enough to start writing this book. The second author was naturally got infected the passion of the first. Thanks are also due to Ms. L.-P. Ding and Mr. M. Toda for their devoted endeavor, without their enthusiastic help, the book would have not been risen out.

the authors

Contents

Chapter 1

The theory of Bernoulli and allied polynomials

Abstract

In this chapter we shall develop the theory of Bernoulli polynomials in a way different from many existing books in that we shall reveal the relationships between the Propositions (D')-(H) and that any one of them (or a combination thereof) can be adopted as a definition of Bernoulli polynomials (cf. Fig. 1. 5 for loose equivalence). Our intension is not to provide a proof of the exact equivalence but equivalence in a loose sense (e.g. up to the initial condition or the normalization) so that the reader can have a better grasp of the formulas scattered around the literature ([Böh], [Ca], [Erd], [Ni]). We shall also state some facts about the cyclotomic polynomials (used in Chapter 2.)

We adopt Lehmer's terminology [Leh2].

Definition 1.1 The **Bernoulli polynomial** $B_n(x)$ of degree n ($n = 0, 1, 2, \ldots$) can be defined by either of the following defining conditions.

(D') (Appell sequence 1832)

$$B_n'(x) = n\, B_{n-1}(x) \tag{1.1}$$

with initial value $B_0(x) = 1$ and with normalization

$$\int_0^1 B_n(x)\, \mathrm{d}x = 0 \quad (n \in \mathbb{N}).$$

If we know the differentiation formula (D'), it is immediate to calculate the k-th derivative:

1

Fig. 1.1   Jacob Bernoulli

$$B_n^{(k)}(x) = \frac{n!}{(n-k)!} B_{n-k}(x). \tag{1.2}$$

By (1.2) we have the Taylor expansion

$$B_n(x) = (B+x)^n = \sum_{k=0}^{n} \binom{n}{k} B_{n-k}(0)\, x^k,$$

and the normalization condition amounts to the recurrence

$$\sum_{k=0}^{n-1} \binom{n}{k} B_k(0) = 0, \quad n \geq 2. \tag{1.3}$$

We denote the value $B_n(0)$ by B_n and refer to it as the n-th **Bernoulli number**. Throughout in what follows we understand they are defined by (1.3) once and for all. Another definition by (1.6) leads to the same recurrence as (1.3) and these two definitions are consistent (cf. Remark 1.1).

(A) Addition formula

$$B_n(x+y) = \sum_{k=0}^{n} \binom{n}{k} B_{n-k}(x)\, y^k$$

$$= \sum_{k=0}^{n} \binom{n}{k} B_{n-k}(y)\, x^k.$$

(U) Umbral calculus formula (Lucas 1891)
The n-th Bernoulli polynomial can be expressed as $(B+x)^n$:

$$B_n(x) = (B+x)^n = \sum_{k=0}^{n} \binom{n}{k} B_{n-k}\, x^k, \tag{1.4}$$

i.e. (1.7), where, by umbral calculus, we mean that after expanding the binomial, the exponent of B is to be degraded to subscript.

Theorem 1.1 *The defining conditions in Definition 1.1 are equivalent to conditions (DE)-(H).*

(DE) $\{B_n(x)\}$ are (principal) solutions of the difference equation

$$\Delta B_n(x) = B_n(x+1) - B_n(x) = n\, x^{n-1},$$

where Δ signifies the difference operator $\Delta u(x) = u(x+1) - u(x)$.

(G) Generating functionology (Euler 1738).
Bernoulli polynomials $B_n(x)$ are defined as the Taylor coefficients of the generating function

$$\frac{z\, e^{xz}}{e^z - 1} = \sum_{n=0}^{\infty} \frac{B_n(x)}{n!}\, z^n \quad (|z| < 2\pi), \tag{1.5}$$

whence, in particular, $B_n = B_n(0)$ are defined by the generating function

$$\frac{z}{e^z - 1} = \sum_{n=0}^{\infty} \frac{B_n}{n!}\, z^n \quad (|z| < 2\pi). \tag{1.6}$$

As will be indicated in Remark 1.1, it follows from (1.5) and (1.6) that $B_n(x)$ is a polynomial of degree n given by

$$B_n(x) = \sum_{k=0}^{n} \binom{n}{k} B_{n-k}\, x^k. \tag{1.7}$$

(K) Kubert identity (Raabe 1851)

$B_n(x)$ is a monic polynomial of degree n satisfying

$$B_n(x) = m^{n-1} \sum_{k=0}^{m-1} B_n\left(\frac{x+k}{m}\right), \qquad (1.8)$$

for each $m \in \mathbb{N}$ and $x \in \mathbb{R}$. This identity is often referred to as the distribution property or the multiplication formula.

(H) Fourier series (Hurwitz 1890)

$$\overline{B}_n(x) = -\frac{n!}{(2\pi i)^n} \sum_{\substack{k=-\infty \\ k \neq 0}}^{\infty} \frac{e^{2\pi i k x}}{k^n}, \qquad (1.9)$$

where $\overline{B}_n(x) = B_n\left(x - [x]\right)$, $[x]$ being the integral part of x, for $n \in \mathbb{N}$ (in the case $n = 1$, we should have (7.9)).

(S) Sums of powers (J. Bernoulli 1705?) As in Comtet [Com] let

$$Z(n, r) = \sum_{k=1}^{n} k^r. \qquad (1.10)$$

Then

$$Z(n, r) = \frac{1}{r+1}\left(B_{r+1}(n+1) - B_{r+1}\right), \quad (r \in \mathbb{N}) \qquad (1.11)$$

and

$$Z(n, r) = \frac{1}{r+1} \sum_{k=0}^{r} \binom{r+1}{k} B_k \cdot (n+1)^{r+1-k}, \quad (r \in \mathbb{N}). \qquad (1.12)$$

This was known to Jacob Bernoulli in his Ars Conjectandi, 1713 (posthumously published; J. Bernoulli died in 1705, and so (S) may be proved in 1705? cf. Lehmer [Leh2]).

Although Formula (1.9) has been known since 1713, numerous papers are still appearing which claim new closed formulas for $Z(n, r)$ up to a certain r, say 1,000.

Remark 1.1   *We note that the function $\dfrac{z}{e^z - 1}$ is analytic in $|z| < 2\pi$ (including the origin cf. Theorem A.8), so that it has the Taylor expansion (1.6), which implies the recurrence (left-hand side of (1.3))$= 0$, $n \geq 2$. $= 1$, $n = 1$, in conformity with (1.3). On the other hand, $\dfrac{z}{e^z - 1} e^{xz}$ has its*

JACOBI BERNOULLI,
Profeff. Bafil. & utriufque Societ. Reg. Scientiar.
Gall. & Pruff. Sodal.
MATHEMATICI CELEBERRIMI,

ARS CONJECTANDI,

OPUS POSTHUMUM.

Accedit

T R A C T A T U S

DE SERIEBUS INFINITIS,

Et EPISTOLA Gallicè fcripta

D E   L U D O   P I L Æ
R E T I C U L A R I S.

B A S I L E Æ,
Impenfis T H U R N I S I O R U M, Fratrum.
cIↃ Iↄcc xiii.

Fig. 1.2   Ars Conjectandi

Taylor expansion in $|z| < 2\pi$, *given by the Cauchy product (or sometimes called Abel convolution)*

$$\frac{z}{e^z - 1}\, e^{xz} = \left(\sum_{k=0}^{\infty} \frac{B_k}{k!} z^k \right) \left(\sum_{l=0}^{\infty} \frac{x^l}{l!} z^l \right),$$

which establishes (1.7) *on comparing the coefficients, i.e.* (G)⇒(U). *The special case* $x = 1$ *gives*

$$B_n(1) = \sum_{k=0}^{n} \binom{n}{k} B_k,$$

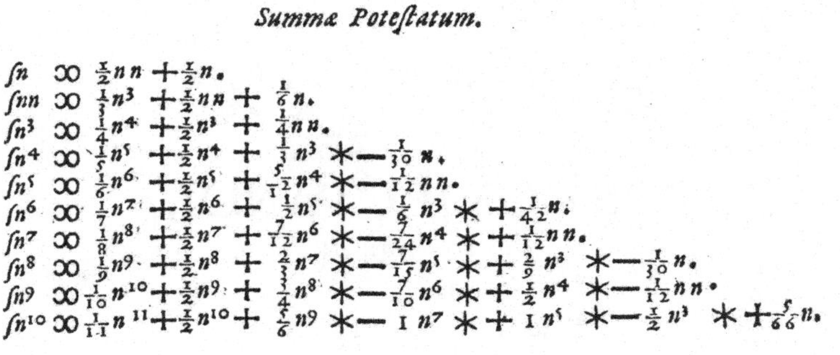

Summæ Potestatum.

$\int n \;\infty\; \frac{1}{2}nn + \frac{1}{2}n.$

$\int nn \;\infty\; \frac{1}{3}n^3 + \frac{1}{2}nn + \frac{1}{6}n.$

$\int n^3 \;\infty\; \frac{1}{4}n^4 + \frac{1}{2}n^3 + \frac{1}{4}nn.$

$\int n^4 \;\infty\; \frac{1}{5}n^5 + \frac{1}{2}n^4 + \frac{1}{3}n^3 * - \frac{1}{30}n.$

$\int n^5 \;\infty\; \frac{1}{6}n^6 + \frac{1}{2}n^5 + \frac{5}{12}n^4 * - \frac{1}{12}nn.$

$\int n^6 \;\infty\; \frac{1}{7}n^7 + \frac{1}{2}n^6 + \frac{1}{2}n^5 * - \frac{1}{6}n^3 * + \frac{1}{42}n.$

$\int n^7 \;\infty\; \frac{1}{8}n^8 + \frac{1}{2}n^7 + \frac{7}{12}n^6 * - \frac{7}{24}n^4 * + \frac{1}{12}nn.$

$\int n^8 \;\infty\; \frac{1}{9}n^9 + \frac{1}{2}n^8 + \frac{2}{3}n^7 * - \frac{7}{15}n^5 * + \frac{2}{9}n^3 * - \frac{1}{30}n.$

$\int n^9 \;\infty\; \frac{1}{10}n^{10} + \frac{1}{2}n^9 + \frac{3}{4}n^8 * - \frac{7}{10}n^6 * + \frac{1}{2}n^4 * - \frac{1}{12}nn.$

$\int n^{10} \;\infty\; \frac{1}{11}n^{11} + \frac{1}{2}n^{10} + \frac{5}{6}n^9 * - 1\,n^7 * + 1\,n^5 * - \frac{1}{2}n^3 * + \frac{5}{66}n.$

Fig. 1.3

which reduces, by (1.3), *to*

$$B_n(1) = B_n(0) = B_n, \quad n \geq 2 \tag{1.13}$$

and $B_1(1) = -B_1$.

Example 1.1 We may calculate Bernoulli numbers from (1.3) or (G): $B_0 = 1$, $B_1 = -\frac{1}{2}$, $B_2 = \frac{1}{6}$, $B_4 = -\frac{1}{30}$, $B_{2k+1} = 0$ $(k \in \mathbb{N})$. The first few Bernoulli polynomials are:

$$\begin{aligned}
B_0(x) &= 1, \; B_1(x) = B_0x + B_1 = x - \frac{1}{2}, \\
B_2(x) &= B_0x^2 + 2B_1x + B_2 = x^2 - x + \frac{1}{6}, \\
B_3(x) &= B_0x^3 + 3B_1x^2 + 3B_2x + B_3 = x^3 - \frac{3}{2}x^2 + \frac{1}{2}x, \\
B_4(x) &= B_0x^4 + 4B_1x^3 + 6B_2x^2 + 4B_3x + B_4 \\
&= x^4 - 2x^3 + x^2 - \frac{1}{30}, \; B_4 = -\frac{1}{30}, \\
B_5(x) &= x^5 - \frac{5}{2}x^4 + \frac{5}{3}x^3 - \frac{1}{6}x, \\
B_6(x) &= x^6 - 3x^5 + \frac{5}{2}x^4 - \frac{1}{2}x^2 + \frac{1}{42}, \; B_6 = \frac{1}{42}.
\end{aligned}$$

We shall prove the equivalence of some of the conditions $(D') \sim (H)$, some being left unproved.

$(D') \Rightarrow (A)$

Indeed, (A) is nothing but the Taylor expansion of $B_n(x+y)$ in y, and the Taylor coefficient is $B_n^{(k)}(x)$, which is $\dfrac{n!}{(n-k)!}B_{n-k}(x)$ by (1.2), whence we have (A).

$(A) \Rightarrow (D')$

For $y \neq 0$ we have

$$\frac{B_n(x+y) - B_n(x)}{y} = \sum_{k=1}^{n} \binom{n}{k} B_{n-k}(x)\, y^{k-1},$$

whence we deduce $B_n'(x) = n\, B_{n-1}(x)$.

$(A) \Rightarrow (U)$

We note that the umbral calculus formula (U) is the special case of (A) with $y = 0$.

$(U) \Rightarrow (A)$

We have by (1.4)

$$\begin{aligned}
B_n(x+y) &= \sum_{k=0}^{n} \binom{n}{k} B_{n-k}(x+y)^k \\
&= \sum_{k=0}^{n} \binom{n}{k} B_{n-k} \sum_{l=0}^{k} \binom{k}{l} y^{k-l} x^l \\
&= \sum_{l=0}^{n} \sum_{k=l}^{n} \binom{n}{k}\binom{k}{l} B_{n-k}\, y^{k-l} x^l,
\end{aligned}$$

where we have changed the order of summation. Applying the formula

$$\binom{n}{k}\binom{k}{l} = \binom{n}{l}\binom{n-l}{k-l}, \tag{1.14}$$

we obtain

$$\begin{aligned}
B_n(x+y) &= \sum_{l=0}^{n} \binom{n}{l} y^l \sum_{k=l}^{n} \binom{n-l}{k-l} B_{n-k}\, x^{k-l} \\
&= \sum_{l=0}^{n} \binom{n}{l} y^l \sum_{k=0}^{n-l} \binom{n-l}{k} B_{n-l-k}\, x^{k}, \tag{1.15}
\end{aligned}$$

by the change of variable. Now the inner sum is $B_{n-l}(x)$ by (1.7).

$(U) \Rightarrow (DE)$ is proved in the following

Exercise 1.1 Deduce (DE) and the reciprocal relation

$$B_n(1 - x) = (-1)^n B_n(x) \tag{1.16}$$

from umbral calculus (U).

Solution By (A), on using (1.15),

$$B_n(1 + y) = \sum_{l=0}^{n} \binom{n}{l} y^l B_{n-l}(1), \tag{1.17}$$

which is $B_n(1 + y) = B_n(y) + n\, y^{n-1}$, by (1.3) and (1.4), and (DE) follows.

By (1.13) and (1.17)

$$B_n(1 - x) = \sum_{l=0}^{n} \binom{n}{l} (-x)^l (-1)^{n-l} B_{n-l}$$

which is $(-1)^n B_n(x)$ by (1.4).

$(U) \Rightarrow (D')$

$$
\begin{aligned}
B_k{}'(x) &= \sum_{r=1}^{k} \binom{k}{r} B_{k-r}\, r\, x^{r-1} \\
&= \sum_{s=0}^{k-1} \binom{k}{s+1} (s+1) B_{k-1-s}\, x^s \\
&= k \sum_{s=0}^{k-1} \binom{k-1}{s} B_{k-1-s}\, x^s \\
&= k B_{k-1}(x).
\end{aligned}
$$

Proof of $\displaystyle\int_0^1 B_n(x)\, dx = 0$ for $n \in \mathbb{N}$.

$$
\begin{aligned}
\int_0^1 B_n(x)\, dx &= \sum_{k=0}^{n} \binom{n}{k} B_{n-k} \int_0^1 x^k\, dx \\
&= \frac{1}{n+1} \sum_{k=0}^{n} \binom{n+1}{k+1} B_{n-k} \\
&= \frac{1}{n+1} \sum_{k=1}^{n+1} \binom{n+1}{k} B_{n+1-k}.
\end{aligned}
$$

Now the sum is $\sum_{k=0}^{n+1} \binom{n+1}{k} B_{n+1-k} - B_{n+1}$, which is $B_{n+1}(1) - B_{n+1}$ by (1.4); this is in turn $B_{n+1}(0) - B_{n+1}$ by (1.16) and is 0. Note that $\int_0^1 B_0(x)\,\mathrm{d}x = 1$.

$(A) \Rightarrow (DE)$
   By (A)

$$B_n(x+1) - B_n(x) = \sum_{k=0}^{n} \binom{n}{k} \left(B_{n-k}(1)\, x^k - B_{n-k}\, x^k \right)$$
$$= \binom{n}{n-1} \left(B_1(1) - B_1 \right) x^{n-1}$$
$$= n\, x^{n-1}$$

on using (1.13).

$(G) \Longleftrightarrow (U)$

$(G) \Rightarrow (U)$ is proved in Remark 1.1.

$(U) \Rightarrow (G)$: We form the generating function $\sum_{n=0}^{\infty} \dfrac{B_n(x)}{n!} z^n$ and substitute (1.4) to get

$$\sum_{n=0}^{\infty} \frac{B_n(x)}{n!} z^n = \sum_{n=0}^{\infty} \frac{1}{n!} \left(\sum_{k+l=n} \frac{n!}{k!\, l!} B_k x^l \right) z^n$$
$$= \left(\sum_{k=0}^{\infty} \frac{B_k}{k!} z^k \right) \left(\sum_{l=0}^{\infty} \frac{x^l}{l!} z^l \right)$$
$$= \frac{z}{e^z - 1} e^{xz},$$

whence (1.5) follows.

$(H) \Rightarrow (K)$
   Substituting (1.9) into the right-hand side of (1.8), we obtain

$$m^{n-1} \sum_{k=0}^{m-1} B_n \left(\frac{x+k}{m} \right) = -\frac{n!}{(2\pi i)^n} \sum_{\substack{r=-\infty \\ r \neq 0}}^{\infty} \frac{e^{2\pi i \frac{x}{m} r}}{r^n} m^n \frac{1}{m} \sum_{k=0}^{m-1} e^{2\pi i \frac{r}{m} k}.$$

The inner sum is 0 except when r is a multiple of m by (8.5). Hence the right-hand side becomes $-\dfrac{n!}{(2\pi i)^n} \displaystyle\sum_{\substack{r=-\infty \\ r\neq 0}}^{\infty} \dfrac{e^{2\pi i x r}}{r^n}$, which is $\overline{B}_n(x)$.

To extend the range $(0, 1)$ to $\mathbb{R}$ we may refer to Milnor's argument [Mi, Lemma 7].

$(G)\Rightarrow(K)$

Consider the generating function

$$\sum_{n=0}^{\infty} \sum_{k=0}^{m-1} B_n\left(\frac{x+k}{m}\right) \frac{(mz)^n}{n!} = \sum_{k=0}^{m-1} \frac{mz\, e^{mz\frac{x+k}{m}}}{e^{mz} - 1}$$

$$= \frac{mz\, e^{xz}}{e^{mz} - 1} \frac{e^{mz} - 1}{e^z - 1} = m\, \frac{z\, e^{xz}}{e^z - 1} = \sum_{n=0}^{\infty} m\, B_n(x)\, \frac{z^n}{n!},$$

whence (1.8) follows.

$(DE)\Rightarrow(S)$

This follows on summing (DE) with $x = 0, 1, \cdots, n$.

Although the above implication is the most natural, we may also apply (G) to deduce (S).

$(G)\Rightarrow(1.11)$

Consider the generating function

$$f_n(z) = \sum_{r=0}^{\infty} \frac{Z(n, r)}{r!} z^{r+1}.$$

On one hand we have

$$f_n(z) = \sum_{r=0}^{\infty} \frac{(r+1)\, Z(n, r)}{(r+1)!} z^{r+1},$$

and on the other, substituting (1.10), we have

$$f_n(z) = z \sum_{r=0}^{\infty} \frac{1}{r!} \sum_{k=1}^{n} k^r z^r = z \sum_{k=1}^{n} \sum_{r=0}^{\infty} \frac{1}{r!} (kz)^r,$$

whence

$$f_n(z) = z \sum_{k=1}^{n} e^{kz}.$$

Rewriting the sum of the geometric series appearing above as

$$z\frac{e^{(n+1)z} - e^z}{e^z - 1} = \frac{ze^{(n+1)z}}{e^z - 1} - \frac{z}{e^z - 1} - z,$$

we have

$$f_n(z) = \sum_{r=0}^{\infty} \frac{z^r}{r!} B_r(n+1) - \sum_{r=0}^{\infty} \frac{z^r}{r!} B_r - z$$

by (G), whence

$$f_n(z) = \sum_{r=1}^{\infty} \left(B_{r+1}(n+1) - B_{r+1} \right) \frac{z^{r+1}}{(r+1)!} + nz;$$

but for $r = 0$:

$$Z(n,0) = B_1(n+1) - B_1 - 1 = n,$$

and so

$$f_n(z) = \sum_{r=0}^{\infty} \left(B_{r+1}(n+1) - B_{r+1} \right) \frac{z^{r+1}}{(r+1)!}.$$

Hence $(G) \Rightarrow (1.11)$ by comparison of the coefficients. (1.12) follows from (1.11) by (U).

We proceed to give another explicit expression for B_r:

$$B_r = \sum_{n=0}^{r} \frac{(-1)^n}{n+1} \binom{r+1}{n+1} Z(n,r). \tag{1.18}$$

We shall deduce (1.18) from (1.12) with the help of (1.19) which gives the closed form for **the Stirling number of the second kind** (cf. (1.20))

$$S(n,k) = \frac{1}{k!} \sum_{j=0}^{k} (-1)^j \binom{k}{j} (k-j)^n \tag{1.19}$$

$$= \frac{1}{k!} \sum_{i=1}^{k} (-1)^{k-i} \binom{k}{i} i^n$$

$$= \frac{1}{k!} \Delta^k 0^n, \quad 1 \leq k \leq n,$$

where Δ is defined in (DE). These are defined as the coefficients in the fundamental relation

$$\sum_{j=1}^{k} S(k,j)(x)_j = x^k = \sum_{j=1}^{k} \Delta^j 0^k \binom{x}{j}, \qquad (1.20)$$

where $(x)_j = x(x-1)\cdots(x-j+1)$ indicates the **falling factorial**. The last equality of (1.19) can be proved as follows:

$$\sum_{j=0}^{k} (-1)^j \binom{k}{j}(k-j)^n = \sum_{j=0}^{k} (-1)^j \binom{k}{j} E^{k-j} 0^n$$
$$= (E-1)^k 0^n$$
$$= \Delta^k 0^n,$$

with $E = \Delta + 1$ the shift operator.

$(1.20) \Rightarrow (1.19)$

Applying the shift operator, $Eu(n) = u(n+1)$, n-times, we obtain

$$n^n = E^n 0^n.$$

The left-hand side is the same as that of (1.20), while on the RHS, we apply $E = \Delta + 1$ formally n-times to deduce that

$$E^n 0^n = (\Delta+1)^n 0^n = \sum_{j=0}^{n} \binom{n}{j} \Delta^j 0^n = \sum_{j=1}^{n} \frac{1}{j!} \Delta^j 0^n (n)_j,$$

whence (1.19) follows.

Substituting (1.12) into the right-hand side of (1.18), we have

$$\text{the RHS} = \sum_{n=0}^{r} \frac{(-1)^n}{n+1} \binom{r+1}{n+1} \frac{1}{r+1} \sum_{k=0}^{r} \binom{r+1}{k} B_k (n+1)^{r+1-k}$$
$$= \frac{1}{r+1} \sum_{k=0}^{r} \binom{r+1}{k} B_k \sum_{n=0}^{r} (-1)^n \binom{r+1}{n+1}(n+1)^{r-k}$$
$$= \frac{1}{r+1} \sum_{k=0}^{r} \binom{r+1}{k} B_k \sum_{j=0}^{r} (-1)^{r+j} \binom{r+1}{j}(r+1-j)^{r-k}$$

on changing the order of summation and then writing $r - n = j$. Divide the sum over k into two: $0 \leq k \leq r-1$ and $k = r$.

$$\text{RHS} = \frac{1}{r+1} \left(\sum_{k=0}^{r-1} \binom{r+1}{k} B_k (-1)^r \sum_{j=0}^{r+1} (-1)^j \binom{r+1}{j} (r+1-j)^{r-k} \right.$$

$$\left. + B_r \left(\sum_{j=0}^{r+1} (-1)^j \binom{r+1}{j} - (-1)^{r+1} \right) \right)$$

$$= \frac{1}{r+1} \sum_{k=0}^{r+1} \binom{r+1}{k} B_k (-1)^r (r+1)! S(r-k, r+1) + (-1)^r B_r,$$

$$= (-1)^r B_r = B_r,$$

where we used the fact that

$$S(n, r) = 0 \quad \text{for } 1 \leq n < r$$

and

$$\sum_{j=0}^{r+1} (-1)^j \binom{r+1}{j} = (1-1)^{r+1} = 0,$$

completing the proof of (1.18).

We shall state an example of (S).

Example 1.2

$$Z(n, 2) = \sum_{k=1}^{n} k^2 = \frac{1}{3} \left(B_3(n+1) - B_3 \right)$$

$$= \frac{1}{3}(n+1) \left\{ (n+1)^2 - \frac{3}{2}(n+1) + \frac{1}{2} \right\}$$

$$= \frac{1}{6}(n+1)(2n^2 + 4n + 2 - 3n - 3 + 1)$$

$$= \frac{1}{6}n(n+1)(2n+1).$$

Compared with (S), the following is less well-known ([Com, p.155]):

Proposition 1.1

$$Z(n, r) = \sum_{j=1}^{r+1} (j-1)! \, S(r+1, j) \binom{n}{j}. \tag{1.21}$$

Proof. By induction, for $n + 1$, the RHS is

$$\sum_{j=1}^{r+1} \frac{1}{j} S(r + 1, j)(n + 1)_j,$$

we rewrite $(n + 1)_j$ as $\dfrac{n + 1 - j + j}{n + 1 - j} = 1 + \dfrac{j}{n + j - 1}$ to obtain

$$\sum_{j=1}^{r+1} \frac{1}{j} S(r + 1, j)(n)_j + \sum_{j=1}^{r+1} S(r + 1, j) n \cdots (n - j + 2),$$

the first term is $Z(n, r)$ by hypothesis and the second can be written as

$$\frac{1}{n + 1} \sum_{j=1}^{r+1} S(r + 1, j)(n + 1)_j,$$

which is $\dfrac{1}{n + 1}(n + 1)^{r+1} = (n + 1)^r$, on applying (1.20). $\square$

The second proof. We first prepare auxiliary results. First,

$$\sum_{j=k}^{n} \binom{j}{k} = \binom{n + 1}{k + 1}, \quad 0 \leq k \leq n. \tag{1.22}$$

This may be proved by writing $(x + 1)^{n+1} - 1$ in two ways: First, it is

$$\frac{(x + 1)^{n+1} - 1}{x} x = \sum_{j=0}^{n} (x + 1)^j x,$$

which becomes

$$\sum_{j=0}^{n} \sum_{k=0}^{j} \binom{j}{k} x^{k+1} = \sum_{k=0}^{n} \sum_{j=k}^{n} \binom{j}{k} x^{k+1}$$

by changing the order of summation. Since $(x + 1)^{n+1} - 1 = \sum_{k=0}^{n} \binom{n+1}{k+1} x^{k+1}$, we obtain (1.22) by comparing the coefficients of x^{k+1}.

Secondly, we also need the triangular recurrence formula for $S(r, j)$:

$$S(r + 1, j) = S(r, j - 1) + j\, S(r, j), \quad 1 \leq j \leq r + 1. \tag{1.23}$$

This may be proved by writing x^{r+1} in (1.20) in two ways:

On one hand, it is $\sum\limits_{j=0}^{r+1} S(r+1,j)(x)_j$, and on the other, it is

$$x \cdot x^r = x \sum_{j=0}^{r} S(r,j)(x)_j.$$

Since

$$x\,(x)_j = (x-j+j)x(x-1)\cdots(x-j+1)$$
$$= (x)_{j+1} + j(x)_j,$$

we have

$$x^{r+1} = \sum_{j=1}^{r} \left(S(r,j-1) + j\,S(r,j)\right)(x)_j,$$

whence (1.23) follows by comparing the coefficients.

We may now prove (1.21). Substituting (1.20), we obtain

$$Z(n,r) = \sum_{j=1}^{r} S(r,j)\,j! \sum_{k=1}^{n}\binom{k}{j}$$

after inverting the order of summation. We rewrite the innermost sum as

$$\binom{n}{j} + \sum_{k=1}^{n-1}\binom{k}{j} = \binom{n}{j} + \binom{n}{j+1}$$

by (1.22). Hence

$$Z(n,r) = \sum_{j=1}^{r+1}\left(S(r,j)\,j!\binom{n}{j} + S(r,j-1)\,(j-1)!\binom{n}{j}\right)$$
$$= \sum_{j=1}^{r+1}\left(j\,S(r,j) + S(r,j-1)\right)(j-1)!\binom{n}{j},$$

which is the RHS of (1.21) in view of (1.23).

Example 1.3 We take up the special case of (1.21).

$$Z(n,2) = \sum_{j=1}^{3}(j-1)!\binom{n}{j}S(3,j).$$

Since from (1.19) we have

$$S(3,1) = 1, \ S(3,2) = 3, \ S(3,3) = 1,$$

it follows that

$$
\begin{aligned}
Z(n,2) &= S(3,1)n + \binom{n}{2}S(3,2) + 2!\binom{n}{3}S(3,3) \\
&= n + \frac{3}{2}n(n-1) + \frac{1}{3}n(n-1)(n-2) \\
&= \frac{1}{6}n(2n^2 + 3n + 1) \\
&= \frac{1}{6}n(n+1)(2n+1).
\end{aligned}
$$

Exercise 1.2   Prove that (H) implies (U) under Euler's identity (5.66).

Solution   Since $B_1(x) = x - \frac{1}{2}$ for $0 < x < 1$, the unique polynomial that coincides with $\overline{B}_1(x)$ is $B_1(x) = x - \frac{1}{2} = x + B_1$. Denoting the right-hand side of (1.9) by $b_n(x)$, we obtain

$$\frac{1}{n}\frac{\mathrm{d}}{\mathrm{d}x} b_n(x) = b_{n-1}(x),$$

whence

$$\frac{1}{n!}\frac{\mathrm{d}^n}{\mathrm{d}x^n} b_n(x) = B_1(x), \quad 0 < x < 1.$$

Integrating $\frac{1}{2}\frac{\mathrm{d}}{\mathrm{d}x}b_2(x) = B_1(x)$, we deduce that

$$\frac{1}{2} b_2(x) = \int B_1(x)\,\mathrm{d}x = \frac{1}{2}x^2 + B_1 x + C,$$

where

$$C = \frac{1}{2} b_2(0) = \frac{1}{2}\frac{2}{\pi^2}\sum_{n=1}^{\infty}\frac{1}{n^2} = \frac{1}{\pi^2}\zeta(2) = \frac{1}{6} = B_2$$

by (5.66).

Repeating this procedure, each time using (5.66), we arrive at (U).

Now we shall follow Lehmer [Leh2] to deduce some of the above defining conditions from (K). First we state a lemma.

Lemma 1.1 (Lehmer)   *For a given $n \in \mathbb{N}$ there is a unique monic polynomial of degree n satisfying (1.8).*

Proof. That $B_n(x)$ satisfies (1.8) is a consequence of (3.69) and (4.1).

To prove uniqueness, suppose there are two polynomials $P_n(x)$ and $Q_n(x)$ of degree n satisfying the conditions. Then

$$R_r(x) := P_n(x) - Q_n(x) = a_0 x^r + a_1 x^{r-1} + \cdots ,$$

where $R_r(x)$ is a polynomial of degree $r < n$ satisfying (1.8):

$$m^{n-1} \sum_{k=0}^{m} R_r \left(\frac{x+k}{m} \right) = R_r(x).$$

Identifying the coefficients of x^r on both sides, we obtain

$$m^n \left(\frac{1}{m} \right)^r a_0 = a_0,$$

which contradicts the fact that $r < n$ and $a_0 \neq 0$.
This completes the proof. $\qquad\square$

Theorem 1.2 (Lehmer) *For $n, m \in \mathbb{N}$, there exists a unique monic polynomial of degree n satisfying the functional equation*

$$\frac{1}{m} \sum_{k=0}^{m-1} f \left(\frac{x+k}{m} \right) = m^{-n} f(x). \tag{1.24}$$

Proof. Since for $m = 1$, (1.24) reduces to a trivial equality, we may assume that $m > 1$.

We substitute a candidate polynomial

$$P_n(x) = b_0 x^n + b_1 x^{n-1} + \cdots + b_n, \quad b_0 \neq 0$$

with b_k as indeterminates, into (1.24). If f satisfies (1.24), then so does any multiple of f, so that we may assume $b_0 = 1$.

The left-hand side of (1.24) becomes

$$\frac{1}{m} \sum_{k=0}^{m-1} P_n \left(\frac{x+k}{m} \right) = \sum_{\nu=0}^{n} b_\nu \sum_{\mu=0}^{n-\nu} x^{n-\nu-\mu} m^{-n+\nu-1} \binom{n-\nu}{\mu} Z(m-1, \mu)$$

$$= m^{-n} \sum_{r=0}^{n} x^{n-r} \sum_{\nu=0}^{r} b_\nu \binom{n-\nu}{r-\nu} m^{\nu-1} Z(m-1, r-\nu).$$

This must be equal to the right-hand side of (1.24):

$$m^{-n} P_n(x) = \sum_{r=0}^{n} x^{n-r} b_r m^{-n}.$$

Identifying the coefficients of x^{n-r}, we conclude that

$$m^{-n}(m^r - 1)b_r = -m^{-n}\sum_{k=0}^{r-1}b_k\binom{n-k}{r-k}m^{k-1}Z(m-1,r-k). \qquad (1.25)$$

Now, from $Z(n,1) = \frac{n(n+1)}{2}$, it follows that

$$b_1 = \binom{n}{1}B_1.$$

Suppose inductively that $b_1, \cdots, b_{r-1}$ $(r > 1)$ are determined. Then (1.25) determines b_r, completing the induction. $\qquad\qquad\square$

We note that by elaborating the above proof, we may actually prove

Proposition 1.2   *The only polynomial satisfying* (1.24) *must be* $B_n(x)$, *on the ground of* (U), (A) *and* (1.16).

Proof.   Indeed, suppose inductively that $b_k = \binom{n}{k}B_k$, $k < r$. Then the right-hand side of (1.25) is

$$-m^{-n}\sum_{k=0}^{r-1}\binom{n}{k}\binom{n-k}{r-k}B_k\,m^{k-1}Z(m-1,r-k).$$

The product of binomial coefficients is $\binom{n}{r}\binom{r}{k}$, and so

$$\text{RHS} = -m^{-n}\binom{n}{r}\sum_{k=0}^{r-1}\binom{r}{k}B_k\,m^{k-1}\frac{1}{r-k+1}\sum_{l=0}^{r-k}\binom{r-k+1}{l}B_l\,m^{r-k+1-l}.$$

Now the innermost sum is

$$m^{r-k+1} + \sum_{l=1}^{r-k}\frac{r-k+1}{l}\binom{r-k}{l-1}B_l\,m^{r-k-(l-1)}.$$

Hence

$$\text{RHS} = -m^{-n+r}\binom{n}{r}S_1 - m^{-m}\binom{n}{r}S_2,$$

where

$$S_1 = \sum_{k=0}^{r-1}\binom{r}{k}B_k\frac{1}{r-k+1} \qquad (1.26)$$

$$S_2 = \sum_{k=0}^{r-1} \binom{r}{k} B_k \sum_{l=1}^{r-k} \frac{1}{l} \binom{r-k}{l-1} B_l \, m^{r-l}. \tag{1.27}$$

Rewriting the product $\binom{r}{k}\binom{r-k}{l-1}$ of binomial coefficients in S_2 by $\binom{r}{l-1}\binom{r-l+1}{r-k-l+1}$, we obtain

$$S_2 = \sum_{l=1}^{r} \frac{1}{l} \binom{r}{l-1} B_l \, m^{r-l} \sum_{k=0}^{r-l} \binom{r-l+1}{r-k-l+1} B_k$$

whose innermost sum is

$$\sum_{k=0}^{r-l} \binom{r-l+1}{k} B_k = \sum_{k=0}^{r-l+1} \binom{r-l+1}{k} B_k - B_{r-l+1}$$

$$= B_{r-l+1}(1) - B_{r-l+1}$$

$$= \begin{cases} 1, & l = r \\ 0, & l \neq r \end{cases}$$

by (U) and (1.16) successively. Since the last member is 0 except when $l = r$, we infer that

$$S_2 = \frac{1}{r} \binom{r}{r-1} B_r = B_r. \tag{1.28}$$

On the other hand,

$$S_1 = \int_0^1 \sum_{k=0}^{r-1} \binom{r}{k} B_k x^{r-k} \, dx = \int_0^1 (B_r(x) - B_r) \, dx$$

$$= \frac{1}{r+1}(B_{r+1}(1) - B_{r+1}(0)) - B_r = -B_r.$$

Hence

$$\text{LHS} = -m^{-n+r} \binom{n}{r}(-B_r) - m^{-n} \binom{n}{r} B_r$$

or

$$m^{-n}(m^r - 1)b_r = -(m^{-n} - m^{-n+r}) \binom{n}{r} B_r,$$

whence

$$b_r = \binom{n}{r} B_r,$$

completing the proof.   □

Exercise 1.3   With $B_n(x)$ as defined by Theorem 1.2, prove (1.5).

Solution   Let

$$F(x,t) = \frac{t\,e^{xt}}{e^t - 1}$$

and expand it into the Taylor series in t:

$$F(x,t) = \sum_{n=0}^{\infty} \frac{1}{n!} G_n(x)t^n, \quad |t| < 2\pi \tag{1.29}$$

with $G_n(x)$ a polynomial of degree n,

$$G_n(x) = a_0^{(n)}x^n + \cdots + a_n^{(n)},$$

say.

We may determine $a_0^{(n)}$ as follows: Replacing x by $\frac{1}{y}$ and t by ty in (1.29), we get

$$\frac{y\,t}{e^{yt} - 1}e^t = \sum_{n=0}^{\infty} \frac{1}{n!} y^n G_n\left(\frac{1}{y}\right) t^n,$$

which leads, as $y \to 0$, to

$$e^t = \sum_{n=0}^{\infty} \frac{1}{n!} a_0^{(n)} t^n$$

(since $y^n G_n\left(\frac{1}{y}\right) \to a_0^{(n)}$):

Hence $a_0^{(n)} = 1$ and G_n is monic.

Now,

$$\frac{1}{m} \sum_{k=0}^{m-1} \sum_{n=0}^{\infty} \frac{1}{n!} G_n\left(\frac{x+k}{m}\right) t^n = \frac{1}{m} \sum_{k=0}^{m-1} F\left(\frac{x+k}{m}, t\right)$$

$$= \frac{1}{m} \frac{t}{e^t - 1} \sum_{k=0}^{m-1} e^{\frac{x+k}{m}t}$$

$$= \frac{1}{m} \frac{t\,e^{\frac{x}{m}t}}{e^{\frac{t}{m}} - 1},$$

which is equal to

$$F\left(x, \frac{t}{m}\right) = \sum_{n=0}^{\infty} \frac{1}{n!} G_n(x) \frac{1}{m^n} t^n.$$

Hence, comparing the coefficients of $\frac{1}{n!}t^n$, we-conclude that

$$\frac{1}{m} \sum_{k=0}^{m-1} G_n\left(\frac{x+k}{m}\right) = m^{-n} G_n(x),$$

whence, by Theorem 1.2, that $G_n(x) = B_n(x)$, completing the proof.

Lemma 1.2 *For every $m \in \mathbb{N}$ there is a unique polynomial satisfying the DE*

$$f(z+1) - f(z) = \sum_{k=1}^{m} \binom{m}{k} b_k z^{m-k}, \quad b_1 \neq 0 \qquad (1.30)$$

with the initial condition $f(1) = 0$.

Proof. From (1.30), $f(z)$ must be of degree m and may be put in the form

$$f(z) = \sum_{j=0}^{m} \binom{m}{j} a_j z^{m-j}, \quad a_0 \neq 0. \qquad (1.31)$$

Forming the difference $f(z+1) - f(z)$, thereby using the expansion

$$(z+1)^{m-j} - z^{m-j} = \sum_{r=1}^{m-j} \binom{m-j}{r} a_j z^{m-j-r}$$

$$= \sum_{k=j+1}^{m} \binom{m-j}{k-j} a_j z^{m-k},$$

we find that

$$f(z+1) - f(z) = \sum_{j=0}^{m} a_j \sum_{k=j+1}^{m} \binom{m}{j}\binom{m-j}{k-j} z^{m-k}.$$

Using (1.14), we get

$$\binom{m}{j}\binom{m-j}{k-j} = \binom{m}{m-j}\binom{m-j}{m-k} = \binom{m}{m-k}\binom{k}{k-j} = \binom{m}{k}\binom{k}{j}.$$

Hence

$$f(z+1) - f(z) = \sum_{k=1}^{m} \binom{m}{k} \left(\sum_{j=0}^{m-1} \binom{k}{j} a_j \right) z^{m-k}. \qquad (1.32)$$

Comparing (1.30) and (1.32), we see that it is enough to show that the following system of $m+1$ linear equations in $m+1$ unknowns $a_0, a_1, \cdots, a_m$ and $f(z)$ has a unique solution:

$$\sum_{j=0}^{k-1} \binom{k}{j} a_j = b_k, \quad k = 1, 2, \cdots, m \qquad (1.33)$$

and

$$-\sum_{j=0}^{m-1} \binom{m}{j} a_j z^{m-j} + f(z) = a_m. \qquad (1.34)$$

This is indeed the case because the coefficient matrix is lower triangular, so that its determinant is the product of all diagonal components $\binom{1}{0}\binom{2}{1}\cdots\binom{m}{m-1} = m! \neq 0$. Hence $f(z)$ is determined by $b_1, \cdots, b_m$ and a_m. Now, comparing (1.33) with $k = m$ and (1.34) with $z = 1$, we find that $a_m = b_m$ by the condition $f(1) = 0$. Hence $f(z)$ is determined uniquely by $b_1, \cdots, b_m$. $\qquad \square$

Theorem 1.3 *For each $m \in \mathbb{N}$ there exists a unique polynomial $f_m(z)$ satisfying the conditions*

$$f_m(z+1) - f_m(z) = m z^{m-1} \qquad (1.35)$$

$$f_m(1) = 0. \qquad (1.36)$$

And the m-th Bernoulli polynomial $B_m(z)$ is defined by

$$B_m(z) = f_m(z) + B_m(1). \qquad (1.37)$$

Proof. This is a special case of Lemma 1.2 with $b_1 = 1$, $b_2 = \cdots = b_m = 0$. $\qquad \square$

Remark 1.2 *As is remarked in [Mi, p.284], the $r+1$-st Bernoulli polynomial can be characterized as the unique polynomial satisfying (1.11) for every natural number r. Hence (S) can be also used as the definition as (K) and (DE).*

Lemma 1.3 *Under (G), Formula (1.37) implies*

$$f'_m(z) = mB_{m-1}(z) = m(f_{m-1}(z) + B_{m-1}), \quad m \in \mathbb{N} \cup \{0\}. \qquad (1.38)$$

Proof. If we differentiate with respect to z

$$\sum_{m=0}^{\infty} \frac{f_m(z)}{m!} x^m = \frac{x}{e^x - 1} e^{xz} - \frac{x e^x}{e^x - 1},$$

we obtain

$$\sum_{m=0}^{\infty} \frac{f'_m(z)}{m!} x^m = \frac{x^2}{e^x - 1} e^{xz} = \sum_{m=0}^{\infty} \frac{B_m(z)}{m!} x^{m+1} = \sum_{m=1}^{\infty} \frac{B_{m-1}(z)}{(m-1)!} x^m,$$

whence we have

$$\frac{1}{m} f'_m(z) = B_{m-1}(z), \quad m \in \mathbb{N},$$

and

$$f'_0(z) = 0,$$

which amounts to (1.38). □

Proposition 1.3 *(DE) and (G) together imply (K).*

Proof. We shall prove that Theorem 1.3 and Lemma 1.3 imply the Kubert identity for $f_m(z)$:

$$f_m(qz) = q^{m-1} \sum_{k=0}^{q-1} f_m\left(z + \frac{k}{q}\right) + (q^{m-1} - 1)B_m(1) \qquad (1.39)$$

for any $q \in \mathbb{N}$.

We follow Böhmer [Böh] to add (1.35) in the form

$$f_{m+1}(z+1) = f_m(z) + (m+1)z^m$$

to the trivial identity

$$q^m \sum_{k=0}^{q-2} f_{m+1}\left(z + \frac{k+1}{q}\right) = q^m \sum_{k=1}^{q-1} f_{m+1}\left(z + \frac{k}{q}\right)$$

to obtain

$$q^m \sum_{k=0}^{q-1} f_{m+1}\left(z + \frac{k+1}{q}\right) = q^m \sum_{k=0}^{q-1} f_{m+1}\left(z + \frac{k}{q}\right) + (m+1)q^m z^m. \qquad (1.40)$$

Now subtract (1.40) from

$$f_{m+1}(qz + 1) = f_{m+1}(qz) + (m + 1)q^m z^m$$

to deduce that

$$
\begin{aligned}
f_{m+1}(qz + 1) - q^m \sum_{k=0}^{q-1} f_{m+1}\left(z + \frac{k+1}{q}\right) & \\
= f_{m+1}(qz) - q^m \sum_{k=0}^{q-1} f_{m+1}\left(z + \frac{k}{q}\right) & \\
= F(z), &
\end{aligned}
\tag{1.41}
$$

say, whence

$$F\left(z + \frac{1}{q}\right) = F(z)$$

i.e. $F(z)$ is periodic of period $\frac{1}{q}$. But, $F(z)$ being a polynomial, we must have $F(z) = $ constant.

Hence differentiating (1.41), we infer that

$$q f'_{m+1}(qz) - q^m \sum_{k=0}^{q-1} f'_{m+1}\left(z + \frac{k}{q}\right) = 0,$$

or

$$q(f_m(qz) + B_m(1)) - q^m \sum_{k=0}^{q-1} \left(f_{m+1}\left(z + \frac{k}{q}\right) + B_m(1)\right) = 0,$$

by Lemma 1.3. This amounts to (1.39), thereby completing the proof. $\square$

Remark 1.3 *Since (DE) and (K) correspond to differentiation and integration, respectively, we need an extra information (i.e. (G)) in Proposition 1.3 to deduce a result on integration from that on differentiation. We also remark that Pan and Sun's result [Su2] implies differentiation from differencing for polynomials, which enables us to prove (D')+(DE)$\Rightarrow$(K). Indeed, differencing the right-hand side of (1.8), we get $\Delta(B_n(x))$, which implies $m^{n-1} \sum_{k=0}^{m-1} \frac{d}{dx} B_n\left(\frac{x+k}{m}\right) = \frac{d}{dx} B_n(x)$, or $m^{n-1} \sum_{k=0}^{m-1} \frac{n}{m} B_{n-1}\left(\frac{x+k}{m}\right) = n B_{n-1}(x)$ by (1.1), whence (1.8) ensues.*

Finally, we add the following as a memorial of the work of L. Lagrange, who was a contemporary of Euler, which is often used in number-theoretic settings.

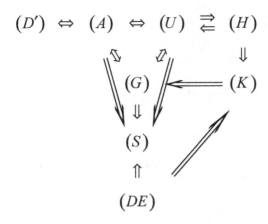

Fig. 1.4   Logical scheme

Fig. 1.5   L. Lagrange

The Lagrange Interpolation Method says that:

Given n values $g(\alpha_i), 1 \leq i \leq n$ of a function $g(x)$ at n distinct arguments α_i, $1 \leq i \leq n$, the Lagrange interpolation polynomial is defined by

$$L(x) = \sum_{i=1}^{n} \frac{f(x)}{(x - \alpha_i) f'(\alpha_i)} g(\alpha_i) \qquad (1.42)$$

interpolates $g(x)$ $(L(\alpha_i) = g(\alpha_i))$, where

$$f(x) = \prod_{i=1}^{n} (x - x_i).$$

Example 1.4 (Carlitz [Car]) Let $k_1, k_2, \cdots, k_n$ be pairwise relatively prime positive integers and let

$$g_i(x) = \frac{x^{k_i} - 1}{x - 1} = \prod_{l=1}^{n-1} \left(x - \zeta_i^l \right), \quad 1 \leq i \leq n,$$

where ζ_i signifies a primitive k_i-th root of 1. Further put

$$f_i(x) = (x - 1) g_i(x) = x^{k_i} - 1 = \prod_{l=1}^{n-1} \left(x - \zeta_i^l \right)$$

and

$$G_i(x) = \prod_{k \neq i} g_k(x), \quad 1 \leq i \leq n. \qquad (1.43)$$

Then the polynomial

$$L_i(x) = \frac{1}{k_i} \sum_{k=1}^{n-1} \frac{f_i(x)}{x - \zeta_i^k} \frac{\zeta_i^k}{G_i(\zeta_i^k)} + \frac{1}{k_i} \phi_i(1) g_i(x) \qquad (1.44)$$

interpolates the polynomial $\phi_i(x)$ of degree $< k_i - 1$ such that

$$\sum_{i=1}^{n} G_i(x) \phi_i(x) = 1.$$

Proof. We note the following facts. Since $f_i'(x) = k_i x^{k_i - 1}$, we have, for a primitive k_i-th root of 1 and $1 \leq l \leq n - 1$,

$$f_i'\left(\zeta_i^l\right) = k_i \zeta_i^{l(k_i - 1)} = k_i \zeta_i^{-l}.$$

Also $G_i(\zeta_i^l) = 0$ for $k \neq i$, $\neq 0$ for $k = i$. Hence by (1.43),

$$G_i(\zeta_i^l)\,\phi_i(\zeta_i^l) = \sum_{k=1}^{n} G_i(\zeta_i^l)\,\phi_i(\zeta_i^l) = 1.$$

Now note that the polynomial

$$L_i(x) = \sum_{k=1}^{n-1} \frac{f_i(x)}{x - \zeta_i^k}\,\frac{\phi_i(\zeta_i^k)}{f_i'(\zeta_i^k)} + \frac{1}{k_i}\,\phi_i(1)\,g_i(x) \qquad (1.45)$$

is the Lagrange interpolation polynomial for $\phi_i(x)$ in view of $g_i(1) = k_i$. From the above data, $f_i'(\zeta_i^k) = k_i\,\zeta_i^{-k}$ and $\phi_i(\zeta_i^k) = \frac{1}{G_i(\zeta_i^k)}$, so that (1.45) transforms into (1.44). □

This identity is essentially used in the proof of the three-term relation of the Dedekind sum, which is a finite sum of the product of two first periodic Bernoulli polynomials.

Example 1.5 Let $f(x) = \prod_{i=1}^{n}(x - \alpha_i)$ be an irreducible polynomial over $\mathbb{Q}$ (α_i being necessarily distinct and called conjugates). Then

$$T_r\left(\frac{f(x)}{x - \alpha}\,\frac{\alpha^r}{f'(\alpha)}\right) = x^r, \qquad 0 \leq r \leq n - 1,$$

where $\alpha = \alpha_1$, say, and T_r means the sum of conjugate elements.

This is because, by the Lagrange interpolation

$$x^r = \sum_{i=1}^{n} \frac{f(x)}{x - \alpha_i}\,\frac{\alpha_i^r}{f'(\alpha_i)}$$

and all summands on the right are conjugate one another, so that the right-hand side is the trace.

This is essentially used to find the dual basis in a finite extension of $\mathbb{Q}$.

Chapter 2

The theory of the gamma and related functions

Abstract

In this chapter we develop the standard theory of the gamma and related functions in a classical fashion starting from the Eulerian integral of the second kind. We follow partly Böhmer [Böh] and Hata [Hata] (cf. also Erdélyi et al [Erd]). Most of the results in this chapter are restated from the zeta-function theoretic point of view in Chapter 5, which can be read parallel to the present chapter.

2.1 Gamma function

First, we develop the theory of the **gamma function** defined by (2.1), one of equivalent conditions to be discussed in Chapter 5. The gamma function, being the Mellin transform of e^{-x} to be mentioned in §7.4, is defined by the **Eulerian integral of the second kind**

$$\Gamma(s) = \int_0^\infty e^{-x} x^{s-1}\, \mathrm{d}x \tag{2.1}$$

for $\sigma > 0$. This improper integral is absolutely and uniformly convergent in the wide sense in $\sigma > 0$, whence it follows that $\Gamma(s)$ is analytic in the right half-plane $\sigma > 0$.

The gamma function may be continued meromorphically over the whole plane by the **difference relation**

$$\Gamma(s) = \frac{\Gamma(s+n+1)}{s(s+1)\cdots(s+n)}, \quad n \in \mathbb{N} \cup \{0\}. \tag{2.2}$$

Fig. 2.1   Euler

Indeed, for $\sigma > 0$, we have by integration by parts,

$$\Gamma(s) = \left[\frac{1}{s}e^{-x}x^s\right]_0^\infty - \frac{1}{s}\int_0^\infty e^{-x}x^s\,\mathrm{d}x = \frac{1}{s}\Gamma(s+1), \qquad (2.3)$$

whose right-hand side is analytic for $\sigma > -1$. From (2.2) we see that $s = -n$, $n \in \mathbb{N} \cup \{0\}$ are simple poles and that the residues at these poles are

$$\operatorname*{Res}_{s=-n} \Gamma(s) = \frac{(-1)^n}{n!}. \qquad (2.4)$$

Also if we put $s = n \in \mathbb{N}$ in

$$\Gamma(s+1) = s\,\Gamma(s), \qquad (2.5)$$

then we have $\Gamma(n+1) = n!$, which means that the gamma function is a function which interpolates the factorial $n!$.

Exercise 2.1 The generalized binomial coefficient is defined by

$$\binom{z}{r} = \frac{(z)_r}{r!} = \frac{\Gamma(z+1)}{\Gamma(r+1)\,\Gamma(z+1-r)} \tag{2.6}$$

for $r \in \mathbb{N} \cup \{0\}$, where $(z)_r = z(z-1)\cdots(z-r+1)$ indicates the **falling factorial**. Show that for $n \in \mathbb{N} \cup \{0\}$

$$\lim_{z \to -n} \binom{z}{r} = (-1)^r \binom{n+r-1}{r} \tag{2.7}$$

and confirm that the usual definition

$$\binom{-n}{r} = (-1)^r \binom{n+r-1}{r}$$

is realized as the limit (2.7).

Solution Writing by (2.2)

$$\frac{\Gamma(z+1)}{\Gamma(z+1-r)} = \frac{\Gamma(z+n+1)}{(z+1)\cdots(z+n)} \frac{(z+1-r)\cdots(z+n)}{\Gamma(z+n+1)}$$
$$= \frac{(z+1-r)\cdots(z+n-1)}{(z+1)\cdots(z+n-1)},$$

we see that

$$\frac{\Gamma(z+1)}{\Gamma(z+1-r)} \to \frac{(-n+1-r)\cdots(-1)}{(-n+1)\cdots(-1)}$$
$$= \frac{(-1)^{n+r-1}\,(n+r-1)!}{(-1)^{n-1}\,(n-1)!}$$

as $z \to -n$. By (2.5), this is

$$(-1)^r \frac{\Gamma(n+r)}{\Gamma(n)} = (-1)^r \binom{n+r-1}{r} r!,$$

and (2.7) follows.

Remark 2.1 *In contrast to $(\lambda)_r$, $\langle\lambda\rangle_r$ indicates the Pochhammer symbol (or the shifted factorial, since $\langle 1 \rangle_r = r!$) defined by*

$$\langle\lambda\rangle_r := \frac{\Gamma(\lambda+r)}{\Gamma(\lambda)} = \begin{cases} 1 & (r=0) \\ \lambda\,(\lambda+1)\cdots(\lambda+r-1) & (r \in \mathbb{N}). \end{cases}$$

The following formula (**Prym's decomposition**) extracts all poles of the gamma function and renders visible its behavior at the poles (cf. (2.4)):

$$\Gamma(s) = \sum_{n=0}^{\infty} \frac{(-1)^n}{n!} \frac{1}{s+n} + \int_1^\infty e^{-x} x^s \, dx. \tag{2.8}$$

The improper integral is absolutely and uniformly convergent in any disc $|s| < R$ and represents an integral function. The proof follows on dividing the interval of integration and noting that

$$\int_0^1 x^{s-1} \sum_{n=0}^{\infty} \frac{(-1)^n x^n}{n!} \, dx = \sum_{n=0}^{\infty} \frac{(-1)^n}{n!} \int_0^1 x^{n+s-1} \, dx = \sum_{n=0}^{\infty} \frac{(-1)^n}{n!} \frac{1}{s+n}.$$

Let the **beta function** $B(\alpha, \beta)$ be defined by the **Eulerian integral of the first kind**

$$B(\alpha, \beta) = \int_0^1 t^{\alpha-1} (1-t)^{\beta-1} \, dt, \quad \operatorname{Re}\alpha > 0, \operatorname{Re}\beta > 0. \tag{2.9}$$

Exercise 2.2 Prove the formula

$$\Gamma(\alpha)\,\Gamma(\beta) = \Gamma(\alpha+\beta)\,B(\alpha,\beta), \tag{2.10}$$

whence in particular

$$\Gamma\left(\frac{1}{2}\right) = \sqrt{\pi}, \tag{2.11}$$

or the value of the probability integral $\int_0^\infty e^{-x^2} \, dx = \frac{\sqrt{\pi}}{2}$.

Solution First, in (2.9), put $t = \sin^2\theta$ to obtain

$$B(\alpha, \beta) = 2 \int_0^{\frac{\pi}{2}} \sin^{2\alpha-1}\theta \, \cos^{2\beta-1}\theta \, d\theta. \tag{2.12}$$

If in (2.1), we put $t = x^2$, then

$$\Gamma(\alpha) = 2 \int_0^\infty x^{2\alpha-1} e^{-x^2} \, dx,$$

whence for $\operatorname{Re}\alpha > 0$, $\operatorname{Re}\beta > 0$,

$$
\begin{aligned}
\Gamma(\alpha)\,\Gamma(\beta) &= 4 \int_0^\infty x^{2\alpha-1} e^{-x^2}\,dx \int_0^\infty y^{2\beta-1} e^{-y^2}\,dy \\
&= 4 \lim_{X\to\infty} \left(\int_0^X x^{2\alpha-1} e^{-x^2}\,dx \int_0^X y^{2\beta-1} e^{-y^2}\,dy \right) \\
&= 4 \lim_{X\to\infty} \int_0^X \int_0^X x^{2\alpha-1} y^{2\beta-1}\, e^{-(x^2+y^2)}\,dx\,dy \\
&= 4 \lim_{X\to\infty} \iint_D x^{2\alpha-1} y^{2\beta-1}\, e^{-(x^2+y^2)}\,dx\,dy, \qquad (2.13)
\end{aligned}
$$

where $D = \left\{ \binom{x}{y} \,\middle|\, 0 \le \sqrt{x^2+y^2} \le X \right\}$. By the change of variable $x = r\cos\theta$, $y = r\sin\theta$, we have the correspondence

$$
D \leftrightarrow \tilde{D} = \left\{ \binom{r}{\theta} \,\middle|\, 0 \le r \le X,\ 0 \le \theta \le \frac{\pi}{2} \right\},
$$

where the absolute value of the Jacobian of this transformation is $\left| \frac{\partial(x,y)}{\partial(r,\theta)} \right| = r$. Hence

$$
\begin{aligned}
\Gamma(\alpha)\,\Gamma(\beta) &= 4 \lim_{X\to\infty} \iint_{\tilde{D}} r^{2\alpha+2\beta-2}\, e^{-r^2} \sin^{2\alpha-1}\theta \cos^{2\beta-1}\theta\, r\,dr d\theta \\
&= 2 \int_0^\infty r^{2\alpha+2\beta-1} e^{-r^2}\,dr \cdot 2 \int_0^{\frac{\pi}{2}} \sin^{2\alpha-1}\theta \cos^{2\beta-1}\theta\,d\theta \\
&= \Gamma(\alpha+\beta)\,\mathrm{B}(\alpha,\beta)
\end{aligned}
$$

by (2.12) above. This completes the proof of (2.10).

Putting now $\alpha = \beta = \frac{1}{2}$, we obtain

$$
\Gamma\!\left(\frac{1}{2}\right)^2 = \Gamma(1)\, 2 \int_0^{\frac{\pi}{2}} d\theta = \pi,
$$

whence (2.11) follows. It follows that

$$
\int_{-\infty}^\infty e^{-\frac{x^2}{2}}\,dx = \sqrt{2\pi},
$$

which is used to normalize the distribution function of the Gaussian (or normal) distribution.

Remark 2.2 *An ordinary procedure for proving* (2.11) *is to use* (2.13) *for $\alpha = \beta = \frac{1}{2}$*

$$\left(\int_0^\infty e^{-x^2}\, dx \right)^2 = 4 \lim_{R \to \infty} \left(\int_0^R r\, e^{-r^2}\, dr \int_0^{\frac{\pi}{2}} d\theta \right) = 4 \left[-\frac{1}{2} e^{-r^2} \right]_0^\infty \frac{\pi}{2} = \pi.$$

Thus we see that if we generalize the problem by introducing the parameters α and β, we get a wider perspective.

Exercise 2.3 Use the formula in Corollary A.4 to prove that

$$\mathrm{B}(s, 1-s) = \frac{\pi}{\sin \pi s} \tag{2.14}$$

or

$$\Gamma(s)\,\Gamma(1-s) = \frac{\pi}{\sin \pi s}, \tag{2.15}$$

the reciprocity relation for the gamma function, where $0 < s < 1$ in the first place and then for all s except for integer values by analytic continuation.

Solution Make the substitution $t = \frac{1}{x+1}$ in (2.9) to deduce that

$$\mathrm{B}(s, 1-s) = \int_0^\infty \frac{x^{-s}}{1+x}\, dx,$$

which proves (2.14) in view of Corollary A.4.

Exercise 2.4 Determine the value of the probability integral $\Gamma\left(\frac{1}{2}\right)$ by considering two functions $f(x) = \left(\int_0^x e^{-t^2}\, dt \right)^2$ and $g(x) = \int_0^1 \frac{1}{t^2 + 1} e^{-x^2(t^2+1)}\, dt.$

Solution Recalling the fundamental theorem of calculus, we obtain

$$f'(x) = 2\, e^{-x^2} \int_0^x e^{-t^2}\, dt.$$

We may differentiate $g(x)$ under the integral sign to get

$$g'(x) = \int_0^1 \frac{1}{t^2 + 1} \frac{d}{dx} e^{-x^2(t^2+1)}\, dt$$

$$= -2x \int_0^1 e^{-x^2(t^2+1)}\, dt = -2\, e^{-x^2} \int_0^x e^{-u^2}\, du$$

by the change of variable $u = xt$.

Hence we conclude that

$$f'(x) + g'(x) = 0,$$

whence by the Newton-Leibniz principle (cf. Lemma B.1) that

$$f(x) + g(x) = f(0) + g(0)$$
$$= \int_0^1 \frac{1}{t^2 + 1}\, dt = \left[\arctan(t)\right]_0^1 = \frac{\pi}{4}.$$

Now letting $x \to \infty$, thereby noting that

$$\lim_{x \to \infty} g(x) = \int_0^1 \frac{1}{t^2 + 1} \lim_{x \to \infty} e^{-x^2(t^2+1)}\, dt = 0,$$

we conclude that

$$\lim_{x \to \infty} f(x) = \left(\int_0^\infty e^{-t^2}\, dt\right)^2 = \frac{\pi}{4},$$

i.e.

$$\int_0^\infty e^{-t^2}\, dt = \frac{\sqrt{\pi}}{2}.$$

Exercise 2.5 Use Formula (2.10) in the form

$$B(\alpha, \beta) = \frac{\Gamma(\alpha)\,\Gamma(\beta)}{\Gamma(\alpha + \beta)} \qquad (2.16)$$

and the value of the probability integral $\Gamma\left(\frac{1}{2}\right) = \sqrt{\pi}$ to deduce **Wallis'** **formula**

$$\frac{\pi}{2} = \lim_{n \to \infty} \left(\frac{2 \cdot 4 \cdots (2n)}{1 \cdot 3 \cdots (2n-1)}\right)^2 \frac{1}{2n + 1}. \qquad (2.17)$$

Solution First, by (2.16) and (2.12)

$$\frac{\Gamma\left(\frac{1}{2}\right)\Gamma\left(n + \frac{1}{2}\right)}{\Gamma(n + 1)} = B\left(\frac{1}{2}, n + \frac{1}{2}\right) = 2\int_0^{\frac{\pi}{2}} \cos^{2n}\theta\, d\theta = S_{2n},$$

say, and

$$\frac{\Gamma\left(\frac{1}{2}\right)\Gamma\left(n + 1\right)}{\Gamma\left(n + \frac{3}{2}\right)} = B\left(\frac{1}{2}, n + 1\right) = 2\int_0^{\frac{\pi}{2}} \cos^{2n+1}\theta\, d\theta = S_{2n+1}.$$

Clearly, $0 < S_{2n+1} < S_{2n} < S_{2n-1}$, so that

$$1 < \frac{S_{2n}}{S_{2n+1}} < \frac{S_{2n-1}}{S_{2n+1}} = \frac{n + \frac{1}{2}}{n} = 1 + \frac{1}{2n} \to 1$$

as $n \to \infty$. Hence

$$\lim_{n \to \infty} \frac{S_{2n}}{S_{2n+1}} = 1. \tag{2.18}$$

Since

$$
\begin{aligned}
\frac{S_{2n}}{S_{2n+1}} &= \frac{\Gamma\left(n + \frac{1}{2}\right) \Gamma\left(n + \frac{3}{2}\right)}{\Gamma(n+1)^2} \\
&= \left(\frac{\left(n - \frac{1}{2}\right)\left(n - \frac{3}{2}\right) \cdots \frac{1}{2} \Gamma\left(\frac{1}{2}\right)}{n \cdot (n - 1) \cdots 1} \right)^2 \left(n + \frac{1}{2} \right) \\
&= \left(\frac{(2n - 1)(2n - 3) \cdots 1}{(2n)(2n - 1) \cdots 2} \right)^2 (2n + 1) \frac{\pi}{2}.
\end{aligned}
$$

Formula (2.18) is nothing but Wallis' formula.

Exercise 2.6 Use Wallis' formula (2.17) to prove that

$$\lim_{N \to \infty} \left(\sum_{n=1}^{N} \log n - \left(N + \frac{1}{2} \right) \log N + N \right) = \log \sqrt{2\pi}, \tag{2.19}$$

whence via (5.23) that

$$\zeta'(0) = -\log \sqrt{2\pi}. \tag{2.20}$$

Solution We write from Corollary 5.1

$$\log \Gamma(x) = \left(x - \frac{1}{2} \right) \log x - x + c + o(1) \tag{2.21}$$

and determine the value of c through the asymptotic formula for

$$\log \left(\binom{2n}{n} \right) = \log \frac{\Gamma(2n + 1)}{\Gamma(n + 1)^2}.$$

First, from (2.21),

$$\log \Gamma(2n + 1) - 2 \log \Gamma(n + 1)$$
$$= \left(2n + \frac{1}{2} \right) \log(2n + 1) - (2n + 1) + c$$
$$- (2n + 1) \log(n + 1) + 2(n + 1) - 2c + o(1)$$
$$= (2n + 1) \log \frac{2n + 1}{n + 1} - \frac{1}{2} \log(2n + 1) + 1 - c + o(1).$$

Secondly, since the first term on the right is

$$(2n + 1) \log \frac{2n + 1}{n + 1} = (2n + 1) \log 2 - (2n + 1) \log \frac{2n + 2}{2n + 1}$$
$$= (2n + 1) \log 2 - (2n + 1) \log \left(1 + \frac{1}{2n + 1} \right)$$
$$= \log 2^{2n} + \log 2 - 1 + O \left(\frac{1}{n} \right),$$

we obtain

$$\log \frac{\Gamma(2n + 1)}{\Gamma(n + 1)^2} = \log 2^{2n} - \log \sqrt{2n + 1} + \log 2 - c + o(1), \qquad (2.22)$$

whence

$$\log \left(\binom{2n}{n} \frac{\sqrt{2n + 1}}{2^{2n}} \right) = \log 2 - c + o(1). \qquad (2.23)$$

Since

$$\frac{1}{2^{2n}} \binom{2n}{n} = \frac{2 \cdot 4 \cdots 2n \cdot 1 \cdot 3 \cdots (2n - 1)}{(2^n n!)^2} = \frac{1 \cdot 3 \cdots (2n - 1)}{2 \cdot 4 \cdots 2n},$$

the LHS of (2.23) is $\log \frac{1 \cdot 3 \cdots (2n-1) \sqrt{2n+1}}{2 \cdot 4 \cdots 2n}$, whose limit is $\log \sqrt{\frac{2}{\pi}}$ by Wallis'
formula. Hence

$$c = \log 2 - \log \sqrt{\frac{2}{\pi}} = \log \sqrt{2\pi}.$$

Recalling the more exact form of (2.21), we may deduce Stirling's formula from the above (cf. Corollary 5.1):

$$\log \Gamma(x) = \left(x - \frac{1}{2} \right) \log x - x + \log \sqrt{2\pi} + O \left(\frac{1}{x} \right). \qquad (2.24)$$

Exercise 2.7   Prove that for $x \geq 2$, the highest power of a prime $p \leq x$ that divides $[x]!$ is given by

$$\sum_{l=0}^{[\log x / \log p]} \left[\frac{x}{p^l}\right],$$

where $[y]$ indicates the greatest integer function introduced in Chapter 1.

Solution   First note that $\left[\frac{x}{p^l}\right]$ is the exact number of multiples of p^l between 1 and x. Among them there are $\left[\frac{x}{p^{l+1}}\right]$ numbers that are multiples of p^{l+1}. Hence $\left[\frac{x}{p^l}\right]$ indicates the exponent of p that appears in $p^l(2p^l)\cdots\left[\frac{x}{p^l}\right]p^l$ but not in $p^{l+1}(2p^{l+1})\cdots\left[\frac{x}{p^{l+1}}\right]p^{l+1}$. Hence adding $\left[\frac{x}{p^l}\right]$ over all l up to r such that $p^r \leq x < p^{r+1}$, i.e. $r = \left[\frac{\log x}{\log p}\right]$, then we obtain the exact exponent β of p such that $p^\beta \parallel [x]!$ (p^β divides $[x]!$ but $p^{\beta+1}$ does not).

Exercise 2.8   Let $\pi(x)$ denote the number of primes $\leq x$. Then using a special case of Stirling's formula (Exercise 2.6, (2.24))

$$\log\left(\binom{2n}{n}\right) = \log\frac{2^{2n}}{\sqrt{2n+1}} + \log\sqrt{\frac{2}{\pi}} + O\left(\frac{1}{n}\right), \qquad (2.25)$$

deduce **Chebyshëv's inequalities**

$$\frac{x}{\log x} \ll \pi(x) \ll \frac{x}{\log x}, \qquad (2.26)$$

where $\ll$ is Vinogradov's symbolism, meaning that there are constants $c_1 > 0$, $c_2 > 0$ such that

$$c_1 \frac{x}{\log x} < \pi(x) < c_2 \frac{x}{\log x}.$$

Solution   First we note the inequalities

$$2\left[\frac{x}{2}\right] \leq [x] \leq 2\left[\frac{x}{2}\right] + 1. \qquad (2.27)$$

The first one is trivial while the second one follows from $\frac{x}{2} - 1 < \left[\frac{x}{2}\right]$ or $x < 2\left[\frac{x}{2}\right]+2$. We consider the highest power of a prime $p \leq 2n$ that divides $\binom{2n}{n}$: $p^\beta \parallel \binom{2n}{n}$. From Exercise 2.7, it is given by

$$\beta = \sum_{l=1}^{r}\left(\left[\frac{2n}{p^l}\right] - 2\left[\frac{n}{p^l}\right]\right), \qquad r = \left[\frac{\log 2n}{\log p}\right]. \qquad (2.28)$$

Bearing (2.27) in mind, we see that each summand is 0 or 1, and so $\beta \leq r \leq \frac{\log 2n}{\log p}$. Hence

$$\binom{2n}{n} = \prod_{p \leq 2n} p^r < (2n)^{\pi(2n)r \log p} < (2n)^{\pi(2n)} \log 2n.$$

Now by (2.25), $\binom{2n}{n} \gg 2^n$, so that

$$\log 2^n \leq \log\left(\binom{2n}{n}\right) \leq \pi(2n) \log 2n.$$

Hence

$$\frac{n}{\log 2n} \ll \pi(2n),$$

so that for $x \geq 2$, using (2.27), we obtain

$$\pi(x) \geq \pi\left(2\left[\frac{x}{2}\right]\right) \gg \frac{\left[\frac{x}{2}\right]}{\log 2\left[\frac{x}{2}\right]} \gg \frac{x}{\log x}.$$

To prove the other inequality in (2.26), we recall the remark after (2.28) to the effect that those prime p, $n < p \leq 2n$ do not divide the denominator, and therefore $\binom{2n}{n}$ must be divisible by $\prod_{n < p \leq 2n} p$. Hence

$$n^{\pi(2n) - \pi(n)} \leq \prod_{n < p \leq 2n} p \leq \binom{2n}{n}.$$

We deduce from this, on using (2.25) in the form $\log\left(\binom{2n}{n}\right) \ll n$, that

$$(\pi(2n) - \pi(n)) \log n \ll n,$$

whence that

$$\pi(2n) \log 2n - \pi(n) \log n = (\pi(2n) - \pi(n)) \log n + \pi(2n) \log 2 \ll n$$

in view of $\pi(x) < x$.

Hence we have

$$\pi(x) \log x - \pi\left(\frac{x}{2}\right) \log \frac{x}{2} \ll x. \tag{2.29}$$

Replacing x in (2.29) by $\frac{x}{2^l}$, $l = 0, \cdots, r$, $r = \left[\frac{\log x}{\log 2}\right]$ $(1 \leq \frac{x}{2^r} < 2)$, we obtain

$$\pi\left(\frac{x}{2^l}\right) \log \frac{x}{2^l} - \pi\left(\frac{x}{2^{l+1}}\right) \log \frac{x}{2^{l+1}} \ll \frac{1}{2^l} x. \tag{2.30}$$

Summing (2.30) for $l = 0, 1, \cdots, r$, we conclude that

$$\pi(x) \log x - \pi \left(\frac{x}{2^r} \right) \log \frac{x}{2^r} \ll x \sum_{l=0}^{r} \frac{1}{2^l}$$

so that $\pi(x) \log x \ll x$, completing the proof.

Theorem 2.1 *The integral representation* (2.1) *and* **Euler's interpolation formula (telescoping series)**

$$\lim_{n \to \infty} \frac{\Gamma(n+z)}{\Gamma(n) \, n^z} = 1, \quad 0 < \operatorname{Re} z < 1 \tag{2.31}$$

are equivalent.

Proof. $\Gamma(s)$ being meromorphic over the whole plane, it suffices to consider the case $0 < z = x < 1$.

First we show that (2.1) implies (2.31). In (2.1) with $s = n \in \mathbb{N}$ put $t = nu$ to obtain

$$\int_0^\infty e^{-nu} u^{n-1} \, du = \Gamma(n) \, n^{-n}. \tag{2.32}$$

Similarly, by (2.1) with $s = n + 1$ and (2.5), we have

$$\int_0^\infty e^{-nu} u^n \, du = \Gamma(n) \, n^{-n}. \tag{2.33}$$

Note that

$$\int_0^1 e^{-nt} t^{n-1} \, dt - \int_0^1 e^{-nt} t^n \, dt = \frac{1}{n} \int_0^1 \frac{d}{dt} (e^{-nt} t^n) \, dt = \frac{1}{n} e^{-n}. \tag{2.34}$$

Subtracting (2.34) from (2.32), we deduce that

$$\int_0^1 e^{-nt} t^n \, dt + \int_1^\infty e^{-nt} t^{n-1} \, dt = \Gamma(n) \, n^{-n} - \frac{1}{n} e^{-n}. \tag{2.35}$$

Adding (2.34) to (2.33), we obtain

$$\int_0^1 e^{-nt} t^{n-1} \, dt + \int_1^\infty e^{-nt} t^n \, dt = \Gamma(n) \, n^{-n} + \frac{1}{n} e^{-n}. \tag{2.36}$$

Hence, for $0 < x < 1$

$$\frac{1}{n^{x+n}}\Gamma(x+n) = \int_0^1 + \int_1^\infty$$

$$< \int_0^1 e^{-nt}t^n \, dt + \int_1^\infty e^{-nt}t^{n-1} \, dt$$

$$< \Gamma(n)\, n^{-n} + \frac{1}{n}e^{-n}$$

by (2.36). Similarly,

$$\frac{1}{n^{x+n}}\Gamma(x+n) > \Gamma(n)\, n^{-n} - \frac{1}{n}e^{-n}.$$

Hence it follows that

$$\Gamma(n)\, n^{-n} - \frac{1}{n}e^{-n} < \frac{\Gamma(x+n)}{n^x}\, n^{-n} < \Gamma(n)\, n^{-n} + \frac{1}{n}e^{-n},$$

or

$$1 - \frac{n^{n-1}e^{-n}}{\Gamma(n)} < \frac{\Gamma(x+n)}{\Gamma(n)\, n^x} < 1 + \frac{n^{n-1}e^{-n}}{\Gamma(n)}.$$

Letting $n \to \infty$, we conclude (2.31).

Now we turn to the deduction of (2.1) from (2.31). Putting

$$\Phi_n(x) = \frac{n^x n!}{x(x+1)\cdots(x+n)},$$

we have

$$\lim_{n\to\infty} \Phi_n(x) = \Gamma(x) \tag{2.37}$$

by (2.31). Expanding $\frac{1}{x(x+1)\cdots(x+n)}$ into partial fractions, we obtain

$$\Phi_n(x) = n^x \sum_{k=0}^n (-1)^k \binom{n}{k} \frac{1}{x+k},$$

whose right-hand side can be expressed as

$$n^x \int_0^1 t^{x-1}(1-t)^n \, dt.$$

Hence

$$\Phi_n(x) = \int_0^n t^{x-1}\left(1 - \frac{t}{n}\right)^n \, dt.$$

Now choose $0 < \varepsilon < \frac{1}{x+2}$ and divide the interval $[0, n]$ into $[0, n^\varepsilon] \cup [n^\varepsilon, n]$. For $0 \le t \le n^\varepsilon$ we have

$$n \log \left(1 - \frac{t}{n}\right) = -t + O\left(t^{2\varepsilon - 1}\right),$$

and so

$$\int_0^{n^\varepsilon} t^{x-1}\left(1 - \frac{t}{n}\right)^n \mathrm{d}t = \int_0^{n^\varepsilon} t^{x-1}e^{-t}\,\mathrm{d}t + O\left(n^{\varepsilon(x+2)-1}\right),$$

while for $n^\varepsilon \le t \le n$,

$$n \log \left(1 - \frac{t}{n}\right) \le n \log \left(1 - n^{\varepsilon - 1}\right) \le -c\,n^\varepsilon, \quad c > 0,$$

so that

$$\int_{n^\varepsilon}^n = O\left(n^\varepsilon e^{-cn^\varepsilon}\right).$$

It follows that

$$\Phi(x) = \int_0^{n^\varepsilon} t^{x-1}e^{-t}\,\mathrm{d}t + o(1),$$

which tends to $\int_0^\infty t^{x-1}e^{-t}\,\mathrm{d}t$. Thus from (2.37) we conclude (2.1).   □

Equivalent definitions of Γ other than (2.1) and (2.31) will be given in Chapter 5 ((5.21), (5.27)$'$, (5.49)).

2.2   The Euler digamma function

As will be discussed in Chapter 5, it is simpler to work with the digamma function and then shift to the gamma function. Below we shall provide several important properties of the digamma function. We follow the lines in Böhmer [Böh]

The following exercise is similar in spirit to Proposition 7.1.

Exercise 2.9   Suppose $f(x)$ is continuous and integrable on $\mathbb{R}_+ = (0, \infty)$. Let $F(x) = \int_0^x f(t)\,\mathrm{d}t$. Then for each $n \in \mathbb{N}$ prove that

$$\int_0^\infty f(x)\,e^{-nx}\,\mathrm{d}x = n \int_0^\infty F(x)\,e^{-nx}\,\mathrm{d}x \tag{2.38}$$

and that

$$\lim_{n \to \infty} \int_0^\infty f(x)\, e^{-nx}\, dx = 0. \tag{2.39}$$

Solution By assumption, $\lim_{x \to \infty} F(x)$ exists and so $F(x) = O(1)$ uniformly on $\mathbb{R}_+$. Construct the auxiliary function

$$G(x) = n \int_0^x F(t)\, e^{-nt}\, dt + F(x)\, e^{-nx}.$$

Then $G(0) = F(0) = 0$ and $G(\infty) = n \int_0^\infty F(t)\, e^{-nt}\, dt$. Also

$$G'(x) = n\, F(x)\, e^{-nx} + F'(x)\, e^{-nx} - n\, F(x)\, e^{-nx}$$
$$= f(x)\, e^{-nx}$$

by the fundamental theorem of calculus. Therefore

$$G(x) = \int_0^x f(t)\, e^{-nt}\, dt + G(0) = \int_0^x f(t)\, e^{-nt}\, dt.$$

Hence, in particular

$$\int_0^\infty f(t)\, e^{-nt}\, dt = G(\infty) = n \int_0^\infty F(t)\, e^{-nt}\, dt,$$

which is (2.38).

Now, to prove (2.39), we note that $F(t)$ is differentiable on $\mathbb{R}_+$ and, *a fortiori*, continuous in the right neighborhood of 0. Hence, for any $\varepsilon > 0$, there is a $\delta > 0$ such that

$$|F(t)| = |F(t) - F(0)| < \varepsilon, \; 0 < t < \delta.$$

Now, divide the interval $(0, \infty)$ into $(0, \delta)$ and (δ, ∞) to obtain

$$\left| \int_0^\infty f(t)\, e^{-nt}\, dt \right| = O\left(n \int_0^\delta |F(t)|\, e^{-nt}\, dt \right) + O\left(n \int_\delta^\infty |F(t)|\, e^{-nt}\, dt \right)$$
$$= O\left(\varepsilon [-e^{-nt}]_0^\delta \right) + O\left([-e^{-nt}]_\delta^\infty \right)$$
$$= O(\varepsilon) + O(e^{-n\delta}),$$

which is $O(\varepsilon)$ if n is big enough. This proves (2.39).

Exercise 2.10 Deduce from (2.39) the integral representation for Euler's constant defined by (5.16):

$$\gamma = -\int_0^\infty f(t,1)\,dt, \tag{2.40}$$

where

$$f(t,z) = \frac{e^{-t}}{t} - \frac{e^{-zt}}{1-e^{-t}}. \tag{2.41}$$

Prove the **Gauss' integral representation** for the digamma function ψ defined by (5.18)

$$\psi(z) = \int_0^\infty f(t,z)\,dt = \int_0^\infty \left(\frac{e^{-t}}{t} - \frac{e^{-zt}}{1-e^{-t}}\right)\,dt, \ \ \mathrm{Re}\,z > 0. \tag{2.42}$$

Solution We notice that $\frac{1}{r} = \int_0^\infty e^{-rt}\,dt$, $r \in \mathbb{N}$. Hence

$$\sum_{r=1}^n \frac{1}{r} = \int_0^\infty \sum_{r=1}^n \left(e^{-t}\right)^r\,dt = \int_0^\infty \frac{1-e^{-nt}}{1-e^{-t}}\,dt. \tag{2.43}$$

By Exercise A.1, (A.2), we have also

$$\log(n+1) = \int_0^\infty \frac{1-e^{-nt}}{t}\,e^{-t}\,dt. \tag{2.44}$$

Subtracting (2.44) from (2.43), we obtain

$$\sum_{r=1}^n \frac{1}{r} - \log(n+1) = -\int_0^\infty f(t,1)\left(1-e^{-nt}\right)\,dt$$

$$\longrightarrow -\int_0^\infty f(t,1)\,dt, \ n \to \infty$$

by Exercise 2.9. Hence, by definition, (2.40) follows.

Let

$$F(z) = \int_0^\infty f(t,z)\,dt,$$

the integral being (absolutely) convergent for $\mathrm{Re}\,z > 0$. Then

$$F(z) + \gamma = \int_0^\infty \frac{e^{-t} - e^{-zt}}{1-e^{-t}}\,dt, \tag{2.45}$$

whence

$$F(z+1) - F(z) = \int_0^\infty e^{-zt} \, dt = \frac{1}{z} \tag{2.46}$$

and

$$F(1) + \gamma = 0, \quad F(1) = -\gamma. \tag{2.47}$$

By the telescoping series technique, we obtain

$$F(z) + \gamma = F(z) - F(1)$$

$$= \sum_{k=0}^{n-1} (F(z+k) - F(1+k) - (F(z+k+1) - F(z+k)))$$

$$+ F(z+n) - F(1+n)$$

$$= \sum_{k=0}^{n-1} \left(\frac{1}{1+k} - \frac{1}{z+k} \right) + F(z+n) - F(1+n) \tag{2.48}$$

by (2.46). Hence

$$\lim_{n \to \infty} (F(z+n) - F(1+n)) = 0 \tag{2.49}$$

is a necessary and sufficient condition for the convergence of the (telescoping) series $\sum_{k=0}^\infty \left(\frac{1}{1+k} - \frac{1}{z+k} \right)$.

Since

$$F(z+n) - F(1+n) = \int_0^\infty \frac{e^{-t} - e^{-zt}}{1 - e^{-t}} e^{-nt} \, dt,$$

(2.49) follows on account of Exercise 2.9.

It follows from (2.48) and (2.49) that

$$F(z) = -\gamma + \sum_{k=0}^\infty \left(\frac{1}{1+k} - \frac{1}{z+k} \right),$$

which is the same as the Gaussian representation for ψ (cf. (5.17)), whence (2.42) follows.

Exercise 2.11 Prove **Legendre's (integral) representation**

$$\psi(z) + \gamma = \int_0^1 \frac{1 - x^{z-1}}{1 - x} \, dx, \quad \operatorname{Re} z > 0. \tag{2.50}$$

Use (2.50) to deduce

$$\frac{\pi}{\sin \pi z} = \int_0^\infty \frac{u^{-z}}{1+u} \, du = \int_0^\infty \frac{v^{z-1}}{1+v} \, dv, \qquad (2.51)$$

and

$$\frac{\pi}{\cos \pi z} = \int_0^\infty \frac{u^{z-\frac{1}{2}}}{1+u} \, du = 2 \int_0^\infty \frac{t^{2z}}{1+t^2} \, dt, \quad -\frac{1}{2} < \operatorname{Re} z < \frac{1}{2} \qquad (2.52)$$

(cf. Corollary A.4).

Solution   (2.50) follows from (2.42) by the change of variable $t = -\log x$.
Using (2.50), we have

$$\psi(z) - \psi\left(\frac{1}{2}\right) = \int_0^1 \frac{x^{-\frac{1}{2}} - x^{z-1}}{1-x} \, dx = -\int_1^\infty \frac{t^{-\frac{1}{2}} - t^{-z}}{1-t} \, dt, \ \operatorname{Re} z > 0, \tag{2.53}$$

where the second expression follows from the first by the change of variable
$x = t^{-1}$. Similarly,

$$\psi(1-z) - \psi\left(\frac{1}{2}\right) = \int_0^1 \frac{t^{-\frac{1}{2}} - t^{-z}}{1-t} \, dt = -\int_1^\infty \frac{x^{-\frac{1}{2}} - x^{z-1}}{1-x} \, dx, \ \operatorname{Re} z < 1. \tag{2.54}$$

Subtraction of (2.54) from (2.53) gives

$$\psi(z) - \psi(1-z) = -\int_0^\infty \frac{t^{-\frac{1}{2}} - t^{-z}}{1-t} \, dt = \int_0^\infty \frac{x^{-\frac{1}{2}} - x^{z-1}}{1-x} \, dx, \ 0 < \operatorname{Re} z < 1.$$

If we admit the relation $\psi(z) = (\log \Gamma(z))'$, then from (2.15), we deduce
that

$$\psi(z) - \psi(1-z) = -\pi \cot \pi z, \tag{2.55}$$

an important relation linking the digamma and the trigonometric functions.
Use this to write the above equality as

$$\pi \cot \pi z = \int_0^\infty \frac{t^{-\frac{1}{2}} - t^{-z}}{1-t} \, dt = -\int_0^\infty \frac{x^{-\frac{1}{2}} - x^{z-1}}{1-x} \, dx,$$

which becomes, on writing $t = u^2$, $x = v^2$, $z \leftrightarrow \dfrac{z}{2}$:

$$\frac{\pi}{2} \cot \frac{\pi}{2} z = \int_0^\infty \frac{1 - u^{1-z}}{1-u^2} \, du = -\int_0^\infty \frac{1 - v^{z-1}}{1-v} \, dv, \ 0 < \operatorname{Re} z < 2 \quad (2.56)$$

and

$$\frac{\pi}{2} \tan \frac{\pi}{2} z = -\int_0^\infty \frac{1 - u^{1-z}}{1 - u^2} \, du = \int_0^\infty \frac{1 - v^{z-1}}{1 - v^2} \, dv, \quad -1 < \operatorname{Re} z < 1,$$

(2.57)

by the interchange of z by $1 - z$.

Adding (2.56) and (2.57), we deduce (2.51), whence (2.52) follows by changing z by $\frac{1}{2} - z$. This completes the proof.

We are now in a position to prove the following remarkable result of Gauss (cf. Theorems 8.1 and 8.2):

Theorem 2.2   *For integers* $1 \le p < q$, *we have*

$$\psi\left(\frac{p}{q}\right) = -\gamma - \log q - \frac{\pi}{2} \cot \frac{p}{q} \pi + \sum_{a=1}^{q-1} \cos \frac{2pk}{q} \log 2 \sin \frac{k}{q} \pi.$$

(2.58)

Proof.   We make the change of variable $x = u^k$ in Legendre's formula (2.50) with $z = \frac{p}{q}$ to obtain

$$\psi\left(\frac{p}{q}\right) + \gamma = \int_0^1 f(u) \, du,$$

(2.59)

where

$$f(u) = q \frac{u^{p-1} - u^{q-1}}{u^q - 1}.$$

(2.60)

Let $\varepsilon = e^{2\pi i \frac{1}{q}}$ be the first primitive q-th root of 1. Then the denominator decomposes into

$$(u - 1) \prod_{l=1}^{q-1} \left(u - \varepsilon^l\right),$$

and the factor $u - 1$ cancels that of the denominator. Hence the partial fraction expansion for $f(u)$ is of the form

$$f(u) = \sum_{k=1}^{q-1} \frac{A_k}{u - \varepsilon^k},$$

and

$$A_k = \lim_{u \to \varepsilon^k} (u - \varepsilon^k) f(u)$$

$$= q \frac{(\varepsilon^k)^{p-1} - (\varepsilon^k)^{q-1}}{\prod_{l=0, l \neq k}^{q-1} (\varepsilon^k - \varepsilon^l)} = \frac{q \varepsilon^{-k}}{\prod_{l=0, l \neq k}^{q-1} \varepsilon^k} \frac{\varepsilon^{pk} - 1}{\prod_{l=0, l \neq k}^{q-1} (1 - \varepsilon^{l-k})}$$

$$= \frac{q}{(\varepsilon^q)^k} \frac{\varepsilon^{pk} - 1}{q} = \varepsilon^{pk} - 1,$$

where we used the identity

$$\prod_{l=1}^{q-1} (1 - \varepsilon^l) = q, \tag{2.61}$$

which follows from the decomposition

$$\prod_{l=1}^{q-1} (u - \varepsilon^l) = u^{q-1} + \cdots + 1. \tag{2.62}$$

Hence

$$\psi\left(\frac{p}{q}\right) + \gamma = \sum_{k=1}^{q-1} (\varepsilon^{pk} - 1) \int_0^1 \frac{1}{u - \varepsilon_k} \, du$$

or

$$\psi\left(\frac{p}{q}\right) + \gamma = \sum_{k=1}^{q-1} (\varepsilon^{pk} - 1) \log \frac{1 - \varepsilon^k}{-\varepsilon^k}. \tag{2.63}$$

Noting that

$$\sum_{k=1}^{q-1} \log \frac{1 - \varepsilon^k}{-\varepsilon^k} = \log \frac{\prod_{k=1}^{q-1} (1 - \varepsilon^k)}{\prod_{k=1}^{q-1} (-\varepsilon^k)} = \log \frac{q}{\left(-\varepsilon^{\frac{q}{2}}\right)^{q-1}} = \log q,$$

we may rewrite (2.63) as

$$\psi\left(\frac{p}{q}\right) + \gamma = \sum_{k=1}^{q-1} \varepsilon^{pk} \log \frac{2i \sin \frac{2\pi}{q} k}{\varepsilon^{\frac{k}{2}}} - \log q, \tag{2.64}$$

which becomes for $q - p$ in place of p:

$$\psi\left(1 - \frac{p}{q}\right) + \gamma = \sum_{k=1}^{q-1} \varepsilon^{pk} \log \frac{2i \sin \frac{2\pi}{q} k}{\varepsilon^{\frac{k}{2}}} - \log q. \qquad (2.65)$$

Adding and subtracting (2.64) and (2.65), we deduce that

$$\psi\left(\frac{p}{q}\right) + \psi\left(1 - \frac{p}{q}\right) + 2\gamma = \sum_{k=1}^{q-1} 2 \cos \frac{2\pi p}{q} k \, \log \sin \frac{2\pi}{q} k - 2 \log q \qquad (2.66)$$

and

$$\psi\left(\frac{p}{q}\right) + \psi\left(1 - \frac{p}{q}\right) = \sum_{k=1}^{q-1} 2i \sin \frac{2\pi p}{q} k \left(-\pi i \frac{k}{q}\right),$$

or

$$\pi \cot \pi \frac{p}{q} = \psi\left(\frac{p}{q}\right) - \psi\left(1 - \frac{p}{q}\right) = \frac{2\pi}{q} \sum_{k=1}^{q-1} k \sin \frac{2\pi p}{q} k. \qquad (2.67)$$

Hence adding (2.66) and (2.67), we obtain

$$2\psi\left(\frac{p}{q}\right) + 2\gamma = 2 \sum_{k=1}^{q-1} \cos \frac{2\pi p}{q} k \, \log \sin \frac{2\pi}{q} k - 2 \log q - \pi \cot \pi \frac{p}{q},$$

whence (2.58) follows, and the proof is complete. $\qquad \square$

Chapter 3

The theory of the Hurwitz-Lerch zeta-functions

Abstract

In this chapter we shall give various contributions to the theory of the Hurwitz zeta-function. In §3.2 we shall give integral representations (for the derivatives as well) which give a basis for the theory of the gamma and related functions to be developed in Chapter 5. In §3.3 we shall give a proof of a formula of Ramanujan whose prototype ($\alpha = 1$) was first stated by Ramanujan and elaborated in [KKaY] In §3.4 we shall give two more proofs of the closed formula for the integral of the psi-function, thus recovering the recent result of Espinosa and Moll. Finally, in §3.5 we shall give another proof of the functional equation.

3.1 Introduction

We shall consider the partial sum defined by (3.5) of the Hurwitz zeta-function defined by (3.1) and prove the integral representation which turns out to hold true for ζ itself. The proof as presented here is quite simple, but the result is far-reaching and we may even base the whole theory of the gamma and related functions on our results (Theorem 3.1 and its corollaries). We shall develop this aspect of our theory in Chapter 5. The special feature of Theorem 3.1 is that the derivatives may be computed by differentiating with respect to u and the whole results may be inherited (for more details, cf. the introductory remark at the beginning of §3.2).

In §3.3, we are going to give the sixth proof of the far-reaching formula

Fig. 3.1   Riemann

of Ramanujan. This proof, incorporating the structure of the Hurwitz zeta-function as the principal solution of the difference equation, seems one of the most natural ones.

We give only a simple example. For more summation formulas going far beyond those in [SC] cf. e.g. [KTTY3].

In §3.4, we shall give two more proofs of the closed formula for the integral $\int_0^z t^\lambda \psi(t+a)\,dt$, thus recovering the seemingly most important formula of Espinosa and Moll [EM1]. We also give two enlightening remarks, the latter of which speaks out the relation between Espinosa and Moll's results and Mikolás' results.

In §3.5, we shall sum up the existing proofs of the functional equation (3.67) for the Hurwitz (Lerch) zeta-function and reveal the hierarchical relationship among them, referring to Laurinčikas and Garunkštis [LG] for the Lerch zeta-function aspects. We shall add one more proof of (3.67) based on the Dirac delta-function. Since from the delta-function, we may deduce the Poisson summation formula, we might regard our proof more fundamental.

Thus, we put the existing literature on the Hurwitz zeta-function in their hierarchical and historical perspective, with our recent contributions [KKaY], [KKSY], [KTTY3] as touchstones.

We define the **Hurwitz zeta-function** and its special case, the **Riemann zeta-function**, by Dirichlet series absolutely convergent for $\sigma > 1$ in the first instance. Both are meromorphically continued over the whole plane with a simple pole at $s = 1$ as we shall see below.

$$\zeta(s, a) = \sum_{n=0}^{\infty} \frac{1}{(n + a)^s}, \qquad \text{Re}\, s = \sigma > 1, \quad a > 0 \tag{3.1}$$

$$\zeta(s) = \zeta(s, 1) = \sum_{n=1}^{\infty} \frac{1}{n^s}, \quad \sigma > 1 \tag{3.2}$$

We define the counterpart of the Hurwitz zeta-function, the **Lerch zeta-function** or the **polylogarithm function** by (cf. Proposition B.1)

$$l_s(a) = \sum_{n=1}^{\infty} \frac{e^{2\pi i n a}}{n^s}, \qquad \sigma > 1, \quad a \in \mathbb{R} \ (\text{or } s = 1, 0 < a < 1) \tag{3.3}$$

We note that $\zeta(s, a)$ satisfies the DE

$$\zeta(s, a + 1) - \zeta(s, a) = a^{-s}. \tag{3.4}$$

We shall use the following notation.

$$\Gamma(s) = \int_0^{\infty} e^{-t} t^{s-1} \, \mathrm{d}t, \quad \sigma > 0$$

– the gamma function;

$$\gamma(s, a) = \int_0^a e^{-t} t^{s-1} \, \mathrm{d}t, \quad \Gamma(s, a) = \int_a^{\infty} e^{-t} t^{s-1} \, \mathrm{d}t$$

– the incomplete gamma functions of the 1st and the 2nd kind (cf. (3.53), (3.66)), which satisfy $\gamma(s, a) + \Gamma(s, a) = \Gamma(s)$;

$$\psi(s) = \frac{\Gamma'(s)}{\Gamma(s)} = \frac{\mathrm{d}}{\mathrm{d}s} \log \Gamma(s)$$

– the Euler digamma function or the psi function;

$$H_n = \psi(n+1) - \psi(1) = \psi(n+1) + \gamma = \sum_{k=1}^{n} \frac{1}{k}$$

– the n-th harmonic number, where γ signifies Euler's constant defined by (5.16) (see below as the Laurent constant $\gamma_0(1) = -\psi(1) = \gamma$).

$$B_n(z) = \sum_{k=0}^{n} \binom{n}{k} B_k \, z^{n-k}$$

– the n-th Bernoulli polynomial with B_k the k-th Bernoulli number defined through

$$\frac{z}{e^z - 1} = \sum_{k=0}^{\infty} \frac{B_k}{k!} z^k \qquad |z| < 2\pi$$

$$\overline{B}_n(z) = B_n(\{z\}) = B_n(z - [z]) \quad \text{for } z \in \mathbb{R}$$

– the n-th periodic Bernoulli polynomial, with $[x]$ and $\{x\}$ signifying the integral and fractional parts of x, respectively.

We use the following as known:

$$B_m(z) = -m\,\zeta(1 - m, z), \quad m \in \mathbb{N}, \quad \text{(cf. (4.1))}$$

$$\psi^{(m)}(z) = (-1)^{m+1} m!\,\zeta(m+1, z), \quad m \in \mathbb{N}, \quad \text{(cf. (5.17))}.$$

The Laurent expansion of $\zeta(s, a)$ at $s = 1$ is given by (cf. Corollary 3.3)

$$\zeta(s, a) = \frac{1}{s - 1} - \psi(a) + \sum_{n=1}^{\infty} \frac{(-1)^n \gamma_n(a)}{n!} (s - 1)^n, \quad s \to 1.$$

The addition formula for the Bernoulli polynomial ((A))

$$B_n(x + y) = \sum_{k=0}^{n} \binom{n}{k} B_k(x) \, y^{n-k}.$$

3.2 Integral representations

For complex variables u and a and $x \geq 0$ let

$$L_u(x, a) = \sum_{0 \leq n \leq x} (n + a)^u, \tag{3.5}$$

denote the partial sum of the Hurwitz zeta-function, where for negative values of u, the possible value of n for which $n + a = 0$ is to be excluded.

We shall use the Euler-Maclaurin sum formula (Theorem B.5, i.e. under Appell's (D')) to prove an integral representation for $L_u(x, a)$, which has the following far-reaching features shared by the derivatives $\frac{\partial^k}{\partial u^k} L_u(x, a)$ as well, i.e. all statements about the function in u ($L_u(x, a)$ and $\zeta(-u, a)$) are valid for their derivatives as well in the form of (i) below.

(i) It gives an analytic expression for $L_u(x, a)$, which entails an integral representation for each derivative $\frac{\partial^k}{\partial u^k} L_u(x, a) = \sum_{0 \le n \le x} (n+a)^u \log^k(n+a)$ (the differentiation of the integral being carried out under the integral sign).

(ii) It gives an asymptotic formula for $L_u(x, a)$ in x by estimating the integral trivially, which is feasible for applications in the divisor problems.

(iii) It gives a generic definition of $\zeta(-u, a)$ for $u \ne -1$ (and for $\gamma_0(a) := -\psi(a)$ for $u = -1$).

(iv) It gives an integral representation for the associated Hurwitz zeta-function $\zeta(-u, a)$ (and its derivatives $\frac{\partial^k}{\partial u^k} \zeta(-u, a) = \zeta^{(k)}(-u, a)$) for $u \ne -1$, and for $u = -1$, it gives an analytic expression for the generalized Euler constant $\gamma_k(a)$ (the k-th Laurent coefficient of $\zeta(s, a)$ at $s = 1$), which follows by simply putting $x = 0$ in the integral representation.

(v) The integral representation for $\zeta(s, a)$ (or $\gamma_k(a)$) in (iii) yields an asymptotic formula for the $\zeta(s, a + z)$ in z with Bernoulli polynomial coefficients (Theorem 2 [Kat1]) as given by Theorem 5.2 below.

Convention. We sometime use $\binom{u}{r} r!$ and $\frac{\Gamma(u+1)}{\Gamma(u+1-r)}$ interchangeably, where the former is suited for easier calculation and the latter for expected differentiation with respect to u.

Theorem 3.1 (Integral Representations) *For any $l \in \mathbb{N}$ with $l > \operatorname{Re} u + 1$ and for any $x \ge 0$, we have the integral representation*

$$
\begin{aligned}
L_u(x, a) = & \sum_{r=1}^{l} \frac{\Gamma(u+1)}{\Gamma(u+2-r)} \frac{(-1)^r}{r!} \overline{B}_r(x) (x+a)^{u-r+1} \\
& + \frac{(-1)^l}{l!} \frac{\Gamma(u+1)}{\Gamma(u+1-l)} \int_x^\infty \overline{B}_l(t) (t+a)^{u-l} \, dt \\
& + \begin{cases} \dfrac{1}{u+1} (x+a)^{u+1} + \zeta(-u, a), & u \ne -1, \\ \log(x+a) - \psi(a), & u = -1. \end{cases}
\end{aligned}
\tag{3.6}
$$

Also the asymptotic formula

$$L_u(x,a) = \sum_{r=1}^{l} \frac{(-1)^r}{r} \binom{u}{r-1} \overline{B}_r(x)(x+a)^{u-r+1} + O\left(x^{\mathrm{Re}\,(u)-l}\right)$$

$$+ \begin{cases} \dfrac{1}{u+1}(x+a)^{u+1} + \zeta(-u,a), & u \neq -1 \\ \log(x+a) - \psi(a), & u = -1 \end{cases} \tag{3.7}$$

holds true as $x \to \infty$.

Furthermore, the integral representation

$$\zeta(-u,a) = a^u - \frac{1}{u+1}a^{u+1} - \sum_{r=1}^{l} \frac{(-1)^r}{r}\binom{u}{r-1} B_r\, a^{u-r+1}$$

$$+ (-1)^{l+1}\binom{u}{l}\int_0^{\infty} \overline{B}_l(t)(t+a)^{u-l}\,dt, \tag{3.8}$$

which follows from (3.6) *by putting* $x = 0$, *holds true for all complex* $u \neq -1$, *where* l *can be any natural number subject only to the condition that* $l > \mathrm{Re}\,u + 1$.

Proof.   Since the r-th derivative of $f(t) = (t+a)^u$ is

$$f^{(r)}(t) = \binom{u}{r}r!\,(t+a)^{u-r} = \frac{\Gamma(u+1)}{\Gamma(u+1-r)}(t+a)^{u-r},$$

we see that the terms in the Euler-Maclaurin sum formula (Theorem B.5) with $a = 0$ become

$$\int_0^x f(t)\,dt = \int_0^x (t+a)^u\,dt = \begin{cases} \dfrac{1}{u+1}(x+a)^{u+1} - \dfrac{1}{u+1}a^{u+1}, & u \neq -1 \\ \log(x+a) - \log(a), & u = -1, \end{cases}$$

$$\sum_{r=1}^{l} \frac{(-1)^r}{r!}\left\{\overline{B}_r(x)f^{(r-1)}(x) - \overline{B}_r(0)f^{(r-1)}(0)\right\}$$

$$= \sum_{r=1}^{l} \frac{\Gamma(u+1)}{\Gamma(u+2-r)}\frac{(-1)^r}{r!}\left\{\overline{B}_r(x)(x+a)^{u-r+1} - B_r\, a^{u-r+1}\right\},$$

and

$$\frac{(-1)^{l+1}}{l!}\int_0^x \overline{B}_l(t)f^{(l)}(t)\,dt = \frac{(-1)^{l+1}}{l!}\frac{\Gamma(u+1)}{\Gamma(u+1-l)}\int_0^x \overline{B}_l(t)(t+a)^{u-l}\,dt,$$

respectively. Hence writing $L_u(x, a) = a^u + \sum\limits_{0 < n \leq x} (n + a)^u$, we obtain

$$
L_u(x, a) = a^u +
\begin{cases}
\dfrac{1}{u + 1} (x + a)^{u+1} - \dfrac{1}{u + 1} a^{u+1}, & u \neq -1 \\
\log(x + a) - \log a, & u = -1
\end{cases}
$$

$$
+ \sum_{r=1}^{l} \frac{\Gamma(u + 1)}{\Gamma(u + 2 - r)} \frac{(-1)^r}{r!} \overline{B_r}(x) (x + a)^{u-r+1}
$$

$$
- \sum_{r=1}^{l} \frac{\Gamma(u + 1)}{\Gamma(u + 2 - r)} \frac{(-1)^r}{r!} B_r a^{u-r+1}
$$

$$
+ \frac{(-1)^{l+1}}{l!} \frac{\Gamma(u + 1)}{\Gamma(u + 1 - l)} \int_0^x \overline{B_l}(t) (t + a)^{u-l} \, dt.
$$

(3.9)

Now for any natural number $l > \operatorname{Re} u + 1$, we have by integration by parts,

$$
\int_x^\infty \overline{B_l}(t) (t + a)^{u-l} \, dt
$$

$$
= \left[\frac{\overline{B_{l+1}}(t)}{l + 1} (t + a)^{u-l} \right]_x^\infty - \frac{u - l}{l + 1} \int_x^\infty \overline{B_{l+1}}(t)(t + a)^{u-l-1} \, dt
$$

$$
= O(x^{\operatorname{Re} u - l}) + O\left(\int_x^\infty (t + a)^{u-l-1} \, dt \right) = O(x^{\operatorname{Re} u - l}),
$$

whence it follows that

$$
\int_0^x \overline{B_l}(t)(t + a)^{u-l} \, dt = \int_0^\infty \overline{B_l}(t)(t + a)^{u-l} \, dt - \int_x^\infty \overline{B_l}(t)(t + a)^{u-l} \, dt
$$

$$
= \int_0^\infty \overline{B_l}(t)(t + a)^{u-l} \, dt + O(x^{\operatorname{Re} u - l}).
$$

Hence we may replace the integral in (3.9) by $\int_0^\infty \overline{B_l}(t)(t + a)^{u-l} \, dt +$

$O\left(x^{\operatorname{Re}u-l}\right)$ to obtain

$$
L_u(x, a) = a^u + \begin{cases} \dfrac{1}{u+1}\,(x+a)^{u+1} - \dfrac{1}{u+1}\,a^{u+1}, & u \neq -1 \\[2mm] \log(x+a) - \log a, & u = -1 \end{cases}
$$

$$
+ \sum_{r=1}^{l} \frac{\Gamma(u+1)}{\Gamma(u+2-r)}\,\frac{(-1)^r}{r!}\,\overline{B}_r(x)\,(x+a)^{u-r+1}
$$

$$
- \sum_{r=1}^{l} \frac{\Gamma(u+1)}{\Gamma(u+2-r)}\,\frac{(-1)^r}{r!}\,B_r\,a^{u-r+1}
$$

$$
+ \frac{(-1)^{l+1}}{l!}\,\frac{\Gamma(u+1)}{\Gamma(u+1-l)} \int_0^\infty \overline{B}_l(t)\,(t+a)^{u-l}\,dt + O\left(x^{\operatorname{Re}u-l}\right).
$$

$$(3.10)$$

Now for $\operatorname{Re}u < -1$ we let $x \to \infty$ to obtain for any $l \in \mathbb{N}$,

$$
\zeta(-u, a) = a^u - \frac{1}{u+1}\,a^{u+1} - \sum_{r=1}^{l} \frac{\Gamma(u+1)}{\Gamma(u+2-r)}\,\frac{(-1)^r}{r!}\,B_r\,a^{u-r+1}
$$

$$
+ \frac{(-1)^l}{l!}\,\frac{\Gamma(u+1)}{\Gamma(u+1-l)} \int_x^\infty \overline{B}_l(t)\,(t+a)^{u-l}\,dt,
$$

$$(3.11)$$

which is (3.8).

Now for any $u \neq -1$, take $l \in \mathbb{N}$ such that $l > \operatorname{Re}u + 1$. Then the last integral in (3.11) is absolutely convergent for $\operatorname{Re}u < l - 1$ and represents an analytic function in $\operatorname{Re}u < l - 1$. Substituting (3.11) in (3.10), we deduce (3.6) for $u \neq -1$.

In the case $u = -1$, the Euler-Maclaurin sum formula on the same lines as above (cf. the proof of (5.39)) gives rise to

$$
L_{-1}(x, a) = (-1)^{l+1}\binom{-1}{l} \int_0^\infty \overline{B}_l(t)\,(t+a)^{-1-l}\,dt
$$

$$
- \sum_{r=1}^{l} \binom{-1}{r-1}\frac{1}{r}\,(-1)^r\,B_r\,a^{-r}
$$

$$
+ \sum_{r=1}^{l} \binom{-1}{r-1}\frac{1}{r}\,(-1)^r\,\overline{B}_r(x)\,(x+a)^{-r}
$$

$$
+ (-1)^l\binom{-1}{l} \int_x^\infty \overline{B}_l(t)\,(t+a)^{-1-l}\,dt
$$

$$
+ a^{-1} + \log(x+a) - \log a.
$$

$$(3.12)$$

Now for x large and $l = 1$, we obtain, from (3.12)

$$\sum_{0 \le n \le x} \frac{1}{n+a} - \log(x+a) = f(a) + O(x^{-1}), \qquad (3.13)$$

where $f(a)$ is a function in a only. If we adopt the definition (5.18) of $\psi(a)$, then we must have $f(a) = -\psi(a)$. Noting that from (3.12) with $x \to \infty$

$$f(a) = -\log a + a^{-1} + \sum_{r=1}^{l} \frac{B_r}{r} a^{-r} - \int_0^\infty \overline{B}_l(t)\,(t+a)^{-1-l}\,dt,$$

we have the integral representation for the digamma function

$$\psi(a) = \log a - \frac{1}{2}a^{-1} + \sum_{r=2}^{l} \frac{B_r}{r} a^{-r} + \int_0^\infty \overline{B}_l(t)\,(t+a)^{-1-l}\,dt. \qquad (3.14)$$

Substituting the constant term $-\psi(a)$ in (3.12), we have the integral representation for the partial sum. $\qquad\square$

Since the integrals appearing in Theorem 3.1 are analytic in the region $\operatorname{Re} u < 1 - l$, we may differentiate (3.6) and (3.8) in u there. We state the counterpart of (3.7) as the following corollaries (the counterpart of (3.8) to be read off from them by putting $x = 0$).

Corollary 3.1 *For any complex u and $a > 0$,*

$$\frac{d}{du} L_u(x,a) = \sum_{0 \le n \le x} (n+a)^u \log(n+a) \qquad (3.15)$$

$$= \sum_{r=1}^{l} \frac{(-1)^r}{r!} \overline{B}_r(x)(x+a)^{u-r+1}$$

$$\times \frac{\Gamma(u+1)}{\Gamma(u+2-r)} \{\psi(u+1) - \psi(u+2-r) + \log(x+a)\}$$

$$+ \frac{(-1)^l}{l!} \int_x^\infty \overline{B}_l(t)(t+a)^{u-l}$$

$$\times \frac{\Gamma(u+1)}{\Gamma(u+1-l)} \{\psi(u+1) - \psi(u+1-l) + \log(t+a)\}\,dt$$

$$+ \begin{cases} \dfrac{1}{u+1}(x+a)^{u+1}\log(x+a) - \dfrac{1}{(u+1)^2}(x+a)^{u+1} \\ \qquad\qquad -\zeta'(-u,a), & u \ne -1 \\ \dfrac{1}{2}\{\log(x+a)\}^2 + \gamma_1(a), & u = -1. \end{cases}$$

Corollary 3.2 *For any complex u and $a > 0$,*

$$\frac{\mathrm{d}^2}{\mathrm{d}u^2} L_u(x, a) \tag{3.16}$$

$$= \sum_{r=1}^{l} \frac{(-1)^r}{r!} \overline{B}_r(x)(x+a)^{u-r+1}$$

$$\times \frac{\Gamma(u+1)}{\Gamma(u+2-r)} \Big[\{\psi(u+1) - \psi(u+2-r) + \log(x+a)\}^2$$

$$+ \psi'(u+1) - \psi'(u+2-r) \Big]$$

$$+ \frac{(-1)^l}{l!} \int_x^{\infty} \overline{B}_l(t)(t+a)^{u-l}$$

$$\times \frac{\Gamma(u+1)}{\Gamma(u+1-l)} \Big[\{\psi(u+1) - \psi(u+1-l) + \log(t+a)\}^2$$

$$+ \psi'(u+1) - \psi'(u+1-l) \Big] \, \mathrm{d}t$$

$$+ \begin{cases} \dfrac{(x+a)^{u+1}}{u+1} \{\log(x+a)\}^2 - \dfrac{2(x+a)^{u+1}}{(u+1)^2} \log(x+a) \\ \quad + \dfrac{2(x+a)^{u+1}}{(u+1)^3} + \zeta''(-u, a), & u \neq -1 \\ \dfrac{1}{3} \{\log(x+a)\}^3 + \gamma_2(a), & u = -1. \end{cases}$$

We note that Theorem 3.1 [(3.6), (3.7)] is first obtained by Mellin [Me] by means of the integral transform under his name (§7.4) and is the most informative for $L_u(x, a)$, so are Corollaries 3.1 and 3.2 for $\frac{\partial}{\partial u} L_u(x, a)$ and $\frac{\partial^2}{\partial u^2} L_u(x, a)$, respectively. Formula (3.6) with $l = 1$, $u \neq -1$, $\mathrm{Re}\, u < 0$ appears as a prototype of the "approximate functional equation" in Landau [Lan, 9–10]. Mikolaś [M1] used it with $x = 1$ to obtain (3.8) with $l = 1$. Then he proceeds to deduce (3.8) with $l = 2$ by integration by parts.

Care should be taken in interpreting the coefficients like $\frac{\Gamma(u+1)}{\Gamma(u+1-l)}(\psi(u+1) - \psi(u+1-l))$ when u is a negative integer; it is to be taken as one without singularities (e.g. in deducing (5.20)).

Corollary 3.3 *The k-th Laurent coefficient of the Hurwitz zeta-function (at $s = 1$) is given by $\frac{(-1)^k}{k!} \gamma_k(a)$, where*

$$\gamma_k(a) = \lim_{x \to \infty} \left(\sum_{0 \le n \le x} \frac{\log^k(n+a)}{n+a} - \frac{\log^{k+1}(x+a)}{k+1} \right) \tag{3.17}$$

and $\gamma_k(a)$ admits the integral representation

$$
\begin{aligned}
\gamma_k(a) = {} & \frac{1}{2a} \log^k a - \frac{1}{k+1} \log^{k+1} a \\
& - \int_0^\infty \frac{\overline{B}_1(t)}{(t+a)^2} \left(\log^k(t+a) - k \log^{k+1}(t+a) \right) dt.
\end{aligned}
\tag{3.18}
$$

In particular, $\gamma_0(a) = -\psi(a)$.

Proof. The following is the simplest possible method known. The starting point is Theorem 3.1 with $l = 1$ and $-s$ ($s \neq 1$, $\sigma > 0$) for u:

$$
L_{-s}(x, a) = \frac{(x+a)^{1-s}}{1-s} + \zeta(s, a) - \frac{\overline{B}_1(x)}{(x+a)^s} + s \int_x^\infty \frac{\overline{B}_1(t)}{(t+a)^{s+1}} \, dt. \tag{3.19}
$$

Since both sides of (3.19) are analytic in $\sigma > 0$, we may compute the k-th Taylor coefficient around $s = 1$. The k-th Taylor coefficient of the left-hand side is

$$
\frac{1}{k!} \frac{\partial^k}{\partial s^k} L_{-s}(x, a)\big|_{s=1} = \frac{(-1)^k}{k!} \sum_{0 \leq n \leq x} (n+a)^{-1} \log^k(n+a) \tag{3.20}
$$

and that of the right-hand side is

$$
\begin{aligned}
\frac{(-1)^k}{k!} & \left(\frac{\log^{k+1}(x+a)}{k+1} + \gamma_k(a) - \frac{\overline{B}_1(x)}{x+a} \log^k(x+a) \right. \\
& \left. + \int_x^\infty \frac{\overline{B}_1(t)}{(t+a)^2} \left(\log^k(t+a) - k \log^{k-1}(t+a) \right) dt \right);
\end{aligned}
\tag{3.21}
$$

equating (3.20) and (3.21), we conclude that

$$
\begin{aligned}
\gamma_k(a) = {} & \sum_{0 \leq n \leq x} (n+a)^{-1} \log^k(n+a) \\
& - \frac{\log^{k+1}(x+a)}{k+1} + \frac{\overline{B}_1(x)}{x+a} \log^k(x+a) \\
& - \int_x^\infty \frac{\overline{B}_1(t)}{(t+a)^2} \left(\log^k(t+a) - k \log^{k-1}(t+a) \right) dt.
\end{aligned}
\tag{3.22}
$$

We now note that (3.22), being valid for any $x \geq 0$, implies both (3.17) and (3.18) by letting $x \to \infty$ and $x = 0$ respectively, (cf. Berndt [Ber3]). In the case $k = 0$, we note that (3.13) and (3.14) correspond to (3.17) and (3.18), respectively. $\square$

3.3   A formula of Ramanujan

In this section we are going to give the sixth proof of the fundamental summation formula based on the use of finite differences, which has been applied successfully in recent researches [KKY3], [KKY2].

Theorem 3.2 (Ramanujan)   *For $0 \leq \lambda \in \mathbb{Z}$ and $|z| < |\alpha|$ we have*

$$
\sum_{m=2}^{\infty} \frac{\zeta(m,\alpha)}{m+\lambda} \, z^{m+\lambda} = \sum_{k=0}^{\lambda} \binom{\lambda}{k} \zeta'(-k, \alpha-z) \, z^{\lambda-k}
$$

$$
- \zeta'(-\lambda, \alpha) - \sum_{k=1}^{\lambda} \frac{1}{k} \, \zeta(k-\lambda, \alpha) \, z^k \qquad (3.23)
$$

$$
+ \frac{1}{\lambda+1} \left(\psi(\alpha) - H_\lambda \right) z^{\lambda+1}.
$$

Proof.   Let $\Delta_\alpha f(\alpha) = f(\alpha+1) - f(\alpha)$ be the difference operator (introduced in (DE) in Chapter 1). We apply this to the sum S on the LHS of (3.23) to obtain

$$
\Delta_\alpha S = \Delta_\alpha \sum_{m=2}^{\infty} \frac{\zeta(m,\alpha)}{m+\lambda} z^{m+\lambda} = - \sum_{m=2}^{\infty} \frac{\alpha^{-m}}{m+\lambda} z^{m+\lambda} = -\alpha^\lambda \sum_{m=\lambda+2}^{\infty} \frac{1}{m} \left(\frac{z}{\alpha} \right)^m.
$$

The resulting infinite series is nothing but

$$
- \log \left(1 - \frac{z}{\alpha} \right) - \sum_{m=1}^{\lambda+1} \frac{1}{m} \left(\frac{z}{\alpha} \right)^m,
$$

or

$$
- \left(\log(\alpha - z) - \log \alpha + \sum_{m=1}^{\lambda+1} \frac{1}{m} \left(\frac{z}{\alpha} \right)^m \right),
$$

whence

$$
\Delta_\alpha S = \alpha^\lambda \log(\alpha - z) - \alpha^\lambda \log \alpha + \sum_{m=1}^{\lambda+1} \frac{\alpha^{\lambda-m}}{m} z^m. \qquad (3.24)
$$

Rewriting the first term on the RHS of (3.24) in the form $\sum_{k=0}^{\lambda} \binom{\lambda}{k} z^{\lambda-k} (\alpha-z)^k \log(\alpha-z)$ and telescoping (3.24), thereby noting that

$$
\zeta'(s, \alpha+1) - \zeta'(s, \alpha) = \alpha^{-s} \log \alpha,
$$

we get

$$
\begin{aligned}
S = &\sum_{k=0}^{\lambda} \binom{\lambda}{k} \zeta'(-k, \alpha - z)\, z^{\lambda-k} - \zeta'(-\lambda, \alpha) \\
&- \sum_{k=1}^{\lambda} \frac{1}{k}\, \zeta(k - \lambda, \alpha)\, z^k + \frac{\psi(\alpha)}{\lambda+1}\, z^{\lambda+1} + f(z, \alpha),
\end{aligned}
\tag{3.25}
$$

where $f(z, \alpha)$ is the function satisfying the conditions

$$
\Delta_\alpha f(z, \alpha) = 0
\tag{3.26}
$$

and

$$
f(0, \alpha) = 0.
\tag{3.27}
$$

It remains to determine $f(z, \alpha)$ (to be $-\frac{H_\lambda}{\lambda+1} z^{\lambda+1}$). First note that

$$
\begin{aligned}
\frac{\mathrm{d}}{\mathrm{d}z}\, \zeta'(-k, \alpha - z) &= \frac{\partial}{\partial s}\frac{\partial}{\partial z}\, \zeta(s, \alpha - z) \,|_{s=-k} \\
&= \frac{\partial}{\partial s}\, s\, \zeta(s+1, \alpha - z) \,|_{s=-k} \\
&= \begin{cases} \zeta(1 - k, \alpha - z) - k\, \zeta'(1 - k, \alpha - z), & k \in \mathbb{N} \\ -\psi(\alpha - z), & k = 0. \end{cases}
\end{aligned}
$$

With this in mind, we differentiate (3.25) with respect to z to obtain

$$
\begin{aligned}
\frac{\partial}{\partial z} S = &\sum_{k=0}^{\lambda-1} \binom{\lambda}{k} (\lambda - k)\, \zeta'(-k, \alpha - z)\, z^{\lambda-k-1} \\
&+ \sum_{k=1}^{\lambda} \binom{\lambda}{k} \zeta(1 - k, \alpha - z)\, z^{\lambda-k} - \sum_{k=1}^{\lambda} \binom{\lambda}{k} \zeta'(1 - k, \alpha - z)\, z^{\lambda-k} \\
&- \psi(\alpha - z)\, z^\lambda - \sum_{k=1}^{\lambda} \zeta(k - \lambda, \alpha)\, z^{k-1} + \psi(\alpha)\, z^{-\lambda} + \frac{\partial}{\partial z} f(z, \alpha).
\end{aligned}
\tag{3.28}
$$

We note that the two sums on the RHS of (3.28) containing ζ' cancel each other, while the second sum, say S_2, becomes, in view of the addition formula,

$$S_2 = -\sum_{k=1}^{\lambda} \binom{\lambda}{k} \frac{1}{k} B_k(\alpha - z) z^{\lambda-k} \tag{3.29}$$

$$= -\sum_{k=1}^{\lambda} \binom{\lambda}{k} \frac{z^{\lambda-k}}{k} \sum_{l=1}^{k} \binom{k}{l} B_l(\alpha) (-z)^{k-l} - \sum_{k=1}^{\lambda} \binom{\lambda}{k} \frac{z^{\lambda-k}}{k} (-z)^k$$

$$= S_{2,1} + S_{2,2},$$

say. Using (1.14) and changing the order of summation in $S_{2,1}$, we have

$$S_{2,1} = -\sum_{l=1}^{\lambda} \binom{\lambda}{l} B_l(\alpha) z^{\lambda-l} \sum_{k=0}^{\lambda-l} \binom{\lambda-l}{k} \frac{(-1)^k}{k+l}.$$

Invoking the formula

$$\sum_{k=0}^{K} \binom{K}{k} \frac{(-1)^k}{k+l} = \frac{K!\,\Gamma(l)}{\Gamma(l+K+1)},$$

we deduce that

$$S_{2,1} = -\sum_{l=1}^{\lambda} \frac{B_l(\alpha)}{l} z^{\lambda-l} = \sum_{l=1}^{\lambda} \zeta(1-l,\alpha) z^{\lambda-l} = \sum_{l=1}^{\lambda} \zeta(l-\lambda,\alpha) z^{l-1}. \tag{3.30}$$

For $S_{2,2}$, we use another formula

$$\sum_{k=1}^{\lambda} \binom{\lambda}{k} \frac{(-1)^k}{k} = H_\lambda$$

to obtain

$$S_{2,2} = H_\lambda z^\lambda. \tag{3.31}$$

From (3.29), (3.30) and (3.31) it follows that

$$S_2 = \sum_{l=1}^{\lambda} \zeta(l-\lambda,\alpha) z^{l-1} + H_\lambda z^\lambda. \tag{3.32}$$

Substituting (3.32) in (3.28), we conclude that

$$\frac{\partial}{\partial z} S = -\psi(\alpha - z) z^\lambda + \psi(\alpha) z^\lambda + H_\lambda z^\lambda + \frac{\partial}{\partial z} f(z,\alpha). \tag{3.33}$$

On the other hand, from (3.25) we know that

$$\frac{\partial}{\partial z} S = - \left(\psi(\alpha - z) - \psi(\alpha) \right) z^\lambda. \tag{3.34}$$

Hence, comparing (3.33) and (3.34), we obtain

$$\frac{\partial}{\partial z} f(z, \alpha) = -H_\lambda z^\lambda,$$

whence

$$f(z, \alpha) = -\frac{H_\lambda}{\lambda + 1} z^{\lambda+1} + C.$$

By condition (3.27), $C = 0$, and

$$f(z, \alpha) = -\frac{H_\lambda}{\lambda + 1} z^{\lambda+1}. \tag{3.35}$$

Substitution of (3.35) into (3.25) completes the proof. □

There are enormous amount of formula (e.g. in [SC], where one third is devoted to the statement of such formulas) which are consequences of Theorem 3.2 (cf. [KKY1], [KKY3] and [KTTY3]). We give only a simple example.

Example 3.1 The formula

$$\log \Gamma(a + 1)$$
$$= \left(a + \frac{1}{2} \right) \log \left(a + \frac{1}{2} \right) + a + \frac{1}{2} - \log \sqrt{2\pi} - \sum_{k=1}^\infty \frac{\zeta(2k, a+1)}{2k(2k+1)} \left(\frac{1}{2} \right)^{2k}$$

is first stated by Wilton [Wil1, Eq.(4)] and is a rather special case of Theorem 3.2. As an asymptotic formula in a, this gives the Stirling formula and is a special case of Corollary 5.1.

3.4 Some definite integrals

We shall give two proofs of [EM2], Theorem 4.3, which seems the highest summit of the paper, and coincides with our Corollary 3 (i) [KKY3]; the first proof depends on a modified form of Theorem 3.2, which we state as Lemma 3.1 while the second depends on a more antecedent one, i.e. the intermediate formula toward the proof of Proposition 1 [KKaY, p.10], which we state as Lemma 3.2.

Theorem 3.3 ([KKY3, Corollary 3 (i)]=[EM2, Theorem 4.3]) *For* $\lambda \in \mathbb{N} \cup \{0\}$, *we have*

$$\int_0^z t^\lambda \psi(t+a)\,dt$$

$$= \sum_{k=0}^\lambda \binom{\lambda}{k}(-1)^k z^{\lambda-k}\left\{\zeta'(-k, a+z) - H_k\,\frac{B_{k+1}(a+z)}{k+1}\right\} \tag{3.36}$$

$$- (-1)^\lambda\left\{\zeta'(-\lambda, a) - H_\lambda\,\frac{B_{\lambda+1}(a+z)}{k+1}\right\}.$$

(3.36) should be compared with our previous result ([KKY3, Corollary 3]): (i) For $\lambda \in \mathbb{N} \cup \{0\}$ and $|z| < \alpha$,

$$\int_0^z t^\lambda \psi(\alpha+t)\,dt$$

$$= (-1)^\lambda \sum_{r=0}^\lambda C_\lambda(r,\alpha) \log \Gamma_{r+1}(\alpha+z)/\Gamma_{r+1}(\alpha)$$

$$+ (-1)^\lambda \sum_{l=1}^\lambda (-1)^l\left\{\binom{\lambda}{l}\zeta'(l-\lambda) + \frac{B_{\lambda-l+1}(\alpha)}{l(\lambda-l+1)}\right\}z^l + \frac{z^{\lambda+1}}{\lambda+1}\,H_\lambda.$$

(ii) For $\lambda \in \mathbb{N}$

$$\lambda \int_0^z t^{\lambda-1} \log \Gamma(\alpha+t)\,dt$$

$$= z^\lambda \log \Gamma(\alpha+z) - (-1)^\lambda \sum_{r=0}^\lambda C_\lambda(r,\alpha) \log \Gamma_{r+1}(\alpha+z)/\Gamma_{r+1}(\alpha)$$

$$- (-1)^\lambda \sum_{l=1}^\lambda (-1)^l\left\{\binom{\lambda}{l}\zeta'(l-\lambda) + \frac{B_{\lambda-l+1}(\alpha)}{l(\lambda-l+1)}\right\}z^l - \frac{z^{\lambda+1}}{\lambda+1}\,H_\lambda,$$

where $\Gamma_r(a)$ signifies the multiple gamma function [SC, p.39].

Lemma 3.1 ([KKaY], (9))   *We have*

$$\sum_{m=2}^{\infty} \frac{\zeta(m,\alpha)}{m+\lambda} z^{m+\lambda}$$

$$= \sum_{k=0}^{\lambda} \binom{\lambda}{k} \{\zeta'(-k,\alpha-z) + H_k\,\zeta(-k,\alpha-z)\} z^{\lambda-k} \qquad (3.37)$$

$$- (\zeta'(-\lambda,\alpha) + H_\lambda\,\zeta(-\lambda,\alpha)) + \frac{\psi(\alpha)}{\lambda+1} z^{\lambda+1}.$$

Proof. (First proof of Theorem 3.3) We start from the Taylor expansion $(|z| < \alpha)$

$$\psi(z+\alpha) = \sum_{n=0}^{\infty} \frac{\psi^{(n)}(\alpha)}{n!} z^n = \psi(\alpha) + \sum_{m=2}^{\infty} (-1)^m \zeta(m,\alpha)\, z^{m-1}. \qquad (3.38)$$

Multiplying both sides of (3.38) by z^λ and integrating over $[0,z]$ with respect to z, we deduce that

$$\int_0^z u^\lambda \psi(\alpha+u)\,du \qquad (3.39)$$

$$= \int_0^z u^\lambda \psi(\alpha)\,du + \sum_{m=2}^{\infty} (-1)^m \zeta(m,\alpha) \int_0^z u^{\lambda+m-1}\,du$$

$$= (-1)^\lambda \sum_{m=2}^{\infty} \frac{\zeta(m,\alpha)}{m+\lambda} (-z)^{m+\lambda} + \frac{z^{\lambda+1}}{\lambda+1} \psi(\alpha).$$

Substituting (3.37) with $-z$ in place of z into (3.39), we obtain

$$\int_0^z u^\lambda \psi(\alpha+u)\,du$$

$$= (-1)^\lambda \sum_{k=0}^{\lambda} \binom{\lambda}{k} \{\zeta'(-k,\alpha+z) + H_k\,\zeta(-k,\alpha+z)\} (-z)^{\lambda-k}$$

$$+ \frac{\psi(\alpha)}{\lambda+1} (-1)^\lambda (-z)^{\lambda+1} + \frac{\psi(\alpha)}{\lambda+1} z^{\lambda+1} \qquad (3.40)$$

$$- (-1)^\lambda (\zeta'(-\lambda,\alpha) + H_\lambda\,\zeta(-\lambda,\alpha)),$$

which is (3.36).   □

Lemma 3.2   *For $0 \leq \lambda \in \mathbb{Z}$ we have*

$$(-1)^{\lambda+1} \int_0^z u^\lambda \zeta(s, a + u)\, du$$

$$= \frac{1}{s-1} \sum_{k=0}^\lambda \binom{\lambda}{k} \frac{\Gamma(2-s)\, k!}{\Gamma(k+2-s)} \zeta(s-k-1, a+z)\,(-z)^{\lambda-k} \qquad (3.41)$$

$$- \frac{1}{s-1} \frac{\lambda!\, \Gamma(2-s)}{\Gamma(\lambda-2-s)} \zeta(s-\lambda-1, a).$$

Proof. (Second proof of Theorem 3.3) Subtracting $\frac{(-1)^{\lambda+1}}{s-1} \int_0^z u^\lambda\, du$ from the left-side, and $\frac{(-1)^{\lambda+1}}{s-1} \frac{z^{\lambda+1}}{\lambda+1}$ from the right-side, of (3.41), we deduce that

$$(-1)^{\lambda+1} \int_0^z u^\lambda \left(\zeta(s, a+u) - \frac{1}{s-1} \right) du = \frac{1}{s-1} F(s), \qquad (3.42)$$

where

$$F(s) = \sum_{k=0}^\lambda \binom{\lambda}{k} \frac{\Gamma(2-s)\, k!}{\Gamma(k+2-s)} \zeta(s-k-1, a+z)\,(-z)^{\lambda-k}$$

$$- \lambda! \frac{\Gamma(2-s)}{\Gamma(\lambda+2-s)} \zeta(s-\lambda-1, a) - \frac{(-z)^{\lambda+1}}{\lambda+1}. \qquad (3.43)$$

We are to take the limit of (3.42) as $s \to 1$. For this we first contend that $F(1) = 0$. Indeed,

$$F(1) = - \sum_{k=0}^\lambda \binom{\lambda}{k} \frac{B_{k+1}(a+z)}{k+1}\, (-z)^{\lambda-k} + \frac{B_{\lambda+1}(a)}{\lambda+1} - \frac{(-z)^{\lambda+1}}{\lambda+1}. \qquad (3.44)$$

Rewriting $\binom{\lambda}{k} \frac{1}{k+1}$ as $\frac{1}{\lambda+1} \binom{\lambda+1}{k+1}$ and writing k for $k+1$, we derive from (3.44) that

$$F(1) = - \frac{1}{\lambda+1} \sum_{k=0}^{\lambda+1} \binom{\lambda+1}{k} B_k(a+z)\,(-z)^{\lambda+1-k} + \frac{B_{\lambda+1}(a)}{\lambda+1}, \qquad (3.45)$$

where we incorporated the last term in (3.39) in the first sum of (3.45). Noting that the fist sum of (3.45) is nothing but the expansion of the Bernoulli polynomial $B_{\lambda+1}(a+z-z) = B_{\lambda+1}(a)$, we conclude $F(1) = 0$.

Hence, we may take the limit as $s \to 1$ of (3.37). On the left side we have $(-1)^\lambda \int_0^z u^\lambda \psi(a+u)\, du$ by the Laurent expansion of $\zeta(s, a+u)$, and on the right-side we just differentiate $F(s)$ with respect to s, thereby noting the formula

$$\left(\frac{\Gamma(2-s)}{\Gamma(k+1-s)}\right)' \Bigg|_{s=1}$$

$$= \frac{\Gamma(2-s)}{\Gamma(k+2-s)}\left(-\psi(2-s)+\psi(k+2-s)\right)\Bigg|_{s=1} = \frac{1}{k!}H_k,$$

(3.46)

to obtain

$$F'(1) = \sum_{k=0}^{\lambda}\binom{\lambda}{k}\left\{\zeta'(-k,a+z)+H_k\,\zeta(-k,a+z)\right\}(-z)^{\lambda-k}$$

$$- \left\{\zeta'(-\lambda,a)+H_\lambda\,\zeta(-\lambda,a)\right\},$$

(3.47)

which is equal to $(-1)^\lambda \int_0^z u^\lambda \psi(a+u)\,\mathrm{d}u$. By multiplying by $(-1)^\lambda$ completes the proof. $\quad\square$

Remark 3.1 *In the notation of [EM2, (3.1), (3.28)],*

$$\zeta'(-k,q)+H_k\,\zeta(-k,q) = \frac{1}{k+1}\left((k+1)\zeta'(-k,q)-H_k B_{k+1}(q)\right)$$

$$= (k+1)!\,\psi^{(-k-1)}(q),$$

and our Theorem 3.3 coincides with Theorem 4.3 of Espinosa and Moll.

Remark 3.2 *(i) Espinosa and Moll [EM1] developed the Hurwitz transform*

$$\int_0^1 f(u)\zeta(s,u)\,\mathrm{d}u$$

and deduced several results for special types of $f(u)$ which can be expanded into Fourier series as consequences of their Theorem 2.2, which in turn is a consequence of the "Fourier series":

$$\zeta(s,u) = 2\,\Gamma(1-s)\sum_{n=1}^{\infty}(2\pi n)^{s-1}\sin\left(2\pi nu+\frac{\pi s}{2}\right),$$

(3.48)

or, more commonly known as the Hurwitz formula (cf. (5.56)). We note that Mikolás' [M3] gave the simplest proof of (3.48) as the Fourier series, whereby he computed the Fourier coefficients

$$\int_0^1 \zeta(s,u)\,e^{-2\pi i\nu u}\,\mathrm{d}u = \frac{\Gamma(1-s)}{(2\pi i\nu)^{1-s}},$$

(3.49)

$0 < s < 1$, $0 \neq \nu \notin \mathbb{Z}$. *From* (3.49) *we immediately deduce*

$$\int_0^1 \zeta(s,u)\cos(2\pi\nu u)\,du - i\int_0^1 \zeta(s,u)\sin(2\pi\nu u)\,du$$

$$= \int_0^1 \zeta(s,u)\,e^{-2\pi i\nu u}\,du = \frac{(2\pi)^s \nu^{s-1}}{4\Gamma(s)}\left(\frac{1}{\cos\frac{\pi}{2}s} - i\frac{1}{\sin\frac{\pi}{2}s}\right),$$

whence follows Formulas (2.2) and (2.3) of Espinosa and Moll.

(ii) Espinosa and Moll [EM1] refer to Mikolás' paper [M2] and quote the result

$$\int_0^1 \zeta(1-s,\{aq\})\,\zeta(1-s,\{bq\})\,dq = 2\,\Gamma^2(s)\,\frac{\zeta(2s)}{(2\pi)^{2s}}\left(\frac{(a,b)}{[a,b]}\right)^s$$

$$((a,b) = g.c.d. \text{ of } a \text{ and } b, \text{ and } [a,b] = l.c.m.)$$

for $\operatorname{Re}(1-s) < \frac{1}{2}$. *We note that Mikolás [M1] obtained the result on the basis of Fourier analysis (the Parseval formula):*

$$\int_0^1 \zeta(s,u)\,\zeta(s',u)\,du$$

$$= 2\,(2\pi)^{s+s'-2}\,\Gamma(1-s)\,\Gamma(1-s')\cos\left(\frac{\pi}{2}(s-s')\right)\zeta(2-s-s')$$

for $\max\{0,\operatorname{Re}s\} + \max\{0,\operatorname{Re}s'\} < 1$; *the region of validity wider than that of Espinosa and Moll who have* $s < 0$, $s' < 0$.

This result of Mikolás', combined with our recent developments of the product of zeta-functions [KTY1], may shed some new light on the asymptotic formula for mean square of zeta-functions. In fact, it looks like the region is one of the excluded one in Katsurada [Kat] and Katsurada and Matsumoto [KM]. For recent developments, cf. [Hashimoto] and [KTZ2].

3.5 The functional equation

In [KKSY] statements were made about the proof of the functional equation, or the Hurwitz formula (3.48), for the Hurwitz zeta-function, using the absolutely convergent Fourier series for $\overline{B}_2(t)$ rather than the boundedly convergent Fourier series for $\overline{B}_1(t)$. Meanwhile the book of Laurinčikas and Garunkštis [LG] has appeared which has rich contents about rather wide spectrum of the theory of the Lerch zeta-function $\phi(\xi,a,s) = \sum_{n=0}^{\infty} \frac{e^{2\pi in\xi}}{(n+a)^s}$ (cf. (8.21)), and we can do no better than referring to it regarding various proofs of the functional equation for ϕ. We shall therefore review mostly those papers which were not quoted in [LG].

As mentioned in Remark 3.2, Mikolás [M3] made use of the Fourier series to deduce the functional equation for $\zeta(s, a)$ and in the subsequent paper [M4], he applied the same method to prove the functional equation for $\phi(\xi, a, s)$.

Berndt [Ber3] used the boundedly convergent Fourier series to deduce (3.48), which he further applied to $\phi(\xi, a, s)$ to deduce the functional equation in [Ber4], where he gave another proof for it, which was reproduced by [LG].

Fine [Fine] applied Riemann's second method, i.e. the theta-transformation formula (for θ_3), or what amounts to the same thing, the Poisson summation formula, to prove (3.48), while Apostol [Ap1] deduced (3.48) from the functional equation and the distribution property for $\phi(\xi, a, s)$.

Apostol's paper [Ap2] (cf. also [Ap3]) contains the seemingly most natural proof of the functional equation for $\phi(\xi, a, s)$ based on the transformation formula and the difference equational structure of ϕ.

As has been developed rather fully in [KTY7], the theta-transformation formula or the modular relation a lá Bochner and the functional equation are equivalent. In this respect, Fine and Apostol would lead to Bochner and may be considered as the prototype of manifestation of the zeta-function associated to prehomogeneous vector spaces.

We remark, however, that although in the above mentioned papers, Lipschitz [Li], Lerch [Le], Hurwitz [H] are referred to, but are neither Malmstén [Ma] nor Schlömilch [Sch], who gave the functional equation for some L-functions (the L-function modulo 4, to be precise), nor the paper of Euler. In this regard we must take into account Weil's paper [We], which gives a translation and comments on Eisenstein's copy of Gauss' Disquisitiones, especially the last page (dated 1849) inserted by the binder. On that page, Eisenstein made an "unmotivated" application of the Poisson summation formula to prove the functional equation for $\phi(\xi, a, s)$ from which he deduces that for L-function mod 4. His argument precedes Oberhettinger [Ob] by 107 years in that he uses the Fourier transform

$$\int_0^\infty e^{2\pi i x y} x^{q-1} \, \mathrm{d}x,$$

while Oberhettinger produces the proof by using the Laplace transform. The Fourier transform is also the basis of Mikolás' proof [M3]. Here we may enjoy a happy encounter of some of the greatest unsimultaneous mathematicians of all time, Gauss, Eisenstein and Weil. We are also fascinated by Weil's imagination on the source of Riemann's paper.

We are indebted to Sato's paper [Sa] for this paper of Weil; without Sato's, we may have missed it, and indeed, in no other places, this discovery of Eisenstein has been presented. E.g. in Grosswald, the Lipschitz transformation formula (i.e. the functional equation) is proved by the Poisson summation formula, which is in principle the same as Eisenstein's proof.

Sato's paper (cf. [KTY7] as well) contains a very nice list of functional equations that follow from the theta-transformation formula and some other deep insight.

We can present a high-brow proof using the Fourier series for the Dirac delta function $\delta(s)$ by completing the incomplete gamma functions.

Our starting point is thus the combination of (41) and (43) of [KKSY] (where we write s for $-u$), which we state as (3.65) below. To derive it, we shall make full use of Formula (3.8) with $l = 2$ (even though any $l \geq 2$ will work here):

$$\zeta(-u, a) = -\frac{1}{u+1} a^{u+1} + \frac{1}{2} a^u - \frac{1}{12} u\, a^{u-1}$$
$$- \frac{u(u-1)}{2!} \int_0^\infty \overline{B}_2(t)(t+a)^{u-2}\, dt, \tag{3.50}$$

valid for $\mathrm{Re}\, s < 2$, which coincides with [UN, Formula (24)].

Substituting the absolutely convergent Fourier series (cf. (1.9)):

$$\overline{B}_2(t) = \frac{1}{2\pi^2} \sum_{n=1}^\infty \frac{e^{2\pi int} + e^{-2\pi int}}{n^2} = \frac{1}{\pi^2} \sum_{n=1}^\infty \frac{\cos(2\pi nt)}{n^2} \tag{3.51}$$

in the integral in (3.8) in the same context as in Rademacher [R, p.83], Pan and Pan [PP, p.125], Kanemitsu [Kan], and Ueno and Nishizawa [UN], we infer after simplification that

$$\int_0^\infty \overline{B}_2(t)(t+a)^{u-2}\,dt \tag{3.52}$$

$$= \frac{1}{2\pi^2} \sum_{n=1}^\infty \frac{1}{n^2} \left(e^{-2\pi i n a}(-2\pi i n)^{1-u} \int_{-2\pi i n a}^\infty e^{-x} x^{u-2}\,dx \right.$$

$$\left. + e^{2\pi i n a}(2\pi i n)^{1-u} \int_{2\pi i n a}^\infty e^{-x} x^{u-2}\,dx \right)$$

$$= \frac{1}{2\pi^2} \sum_{n=1}^\infty \frac{1}{n^2} \left\{ e^{-2\pi i n a}(-2\pi i n)^{1-u} \Gamma(u-1,-2\pi i n a) \right.$$

$$\left. + e^{2\pi i n a}(2\pi i n)^{1-u} \Gamma(u-1,2\pi i n a) \right\},$$

where $\Gamma(s,z)$ designates the **incomplete gamma function of the second kind** defined by

$$\Gamma(s,z) = \int_z^\infty e^{-x} x^{s-1}\,dx. \tag{3.53}$$

The function $\Gamma(s,z)$ can be expressed in terms of the confluent hypergeometric function as follows [Erd, p.266, Eq. 6.9 (21)]:

$$\Gamma(s,z) = e^{-z}\,\Psi(1-s,1-s;z), \tag{3.54}$$

where [Erd, p.255, Eq. 6.5(2)]

$$\Psi(a,c;z) = \frac{1}{\Gamma(a)} \int_0^\infty e^{-zt} t^{a-1}(1+t)^{c-a-1}\,dt, \quad \min\{\operatorname{Re} a, \operatorname{Re} z\} > 0, \tag{3.55}$$

is a solution of the differential equation:

$$z\frac{d^2 w}{dz^2} + (c-z)\frac{dw}{dz} - aw = 0 \tag{3.56}$$

and is denoted by $U(a,c;z)$ in Slater [Sla].

Using [Erd, p.257, Eq. 6.5(6)]

$$\Psi(a,c;z) = z^{1-c}\,\Psi(a-c+1,2-c;z) \tag{3.57}$$

in (3.54), we get

$$e^z z^{-s} \Gamma(s,z) = \Psi(1,1+s;z). \tag{3.58}$$

Substituting (3.58) in (3.52), we may write

$$
\int_0^\infty \overline{B}_2(t)(t+a)^{u-2}\,dt
$$
$$
= \frac{1}{2\pi^2} \sum_{n=1}^\infty \frac{1}{n^2}\, a^{u-1}\left\{\Psi(1,u;-2\pi ina)+\Psi(1,u;2\pi ina)\right\}, \tag{3.59}
$$

which, when substituted in (3.50), gives rise to

$$
\zeta(-u,a)
$$
$$
= -\frac{1}{u+1}\, a^{u+1} + \frac{1}{2}\, a^u - \frac{1}{12}\, u\, a^{u-1}
$$
$$
- \frac{u(u-1)}{2}\frac{a^{u-1}}{2\pi^2} \sum_{n=1}^\infty \frac{1}{n^2}\left\{\Psi(1,u;-2\pi ina)+\Psi(1,u;2\pi ina)\right\}. \tag{3.60}
$$

To deduce the Ueno-Nishizawa formula [UN, Formula (25)], we take $l=1$ and argue in the same way. We just state here the corresponding formulas:

$$
\zeta(-u,a) = -\frac{1}{u+1}\, a^{u+1} + \frac{1}{2}\, a^u + u\int_0^\infty \overline{B}_1(t)(t+a)^{u-1}\,dt, \tag{3.61}
$$

$$
\overline{B}_1(t) = -\frac{1}{2\pi i}\sum_{n=1}^\infty \left(\frac{e^{2\pi int}-e^{-2\pi int}}{n}\right) = -\frac{1}{\pi}\sum_{n=1}^\infty \frac{\sin(2\pi nt)}{n}, \tag{3.62}
$$

$$
\int_0^\infty \overline{B}_1(t)(t+a)^{u-1}\,dt
$$
$$
= -\frac{1}{2\pi i}\sum_{n=1}^\infty \frac{a^u}{n}\left\{e^{-2\pi ina}(-2\pi ina)^{-u}\,\Gamma(u,-2\pi ina)\right. \tag{3.63}
$$
$$
\left. - e^{2\pi ina}(2\pi ina)^{-u}\,\Gamma(u,2\pi ina)\right\},
$$

and

$$
\zeta(-u,a) = -\frac{1}{u+1}\, a^{u+1} + \frac{1}{2}\, a^u
$$
$$
- \frac{u\, a^u}{2\pi i}\sum_{n\neq 0}\frac{1}{n}\,\Psi(1,u+1;-2\pi ina), \tag{3.64}
$$

which is [UN, Formula (25)].

We shall use the combination of (3.61) and (3.63):

$$\zeta(s,a) = \frac{1}{a^{s-1}} \sum_{n=1}^{\infty} \left(\frac{e^{-2\pi i n a}}{(-2\pi i n a)^{1-s}} \Gamma(1-s, -2\pi i n a) \right.$$
$$\left. + \frac{e^{2\pi i n a}}{(2\pi i n a)^{1-s}} \Gamma(1-s, 2\pi i n a) \right) \qquad (3.65)$$
$$+ \frac{1}{2a^s} + \frac{1}{a^{s-1}} \frac{1}{s-1}.$$

We use the **incomplete gamma function** $\gamma(s,a)$ **of the first kind**

$$\gamma(s,a) = \int_0^a e^{-u} u^{s-1} \, du = a^s \int_0^1 e^{-au} u^{s-1} \, du \qquad (3.66)$$

and complete $\Gamma(1-s,a)$ to write $\Gamma(1-s,a) = \Gamma(1-s) - \gamma(1-s,a)$. Thus

$$\zeta(s,a) = \Gamma(1-s) \sum_{n=1}^{\infty} \left(\frac{e^{-2\pi i n a}}{(-2\pi i n)^{1-s}} + \frac{e^{2\pi i n a}}{(2\pi i n)^{1-s}} \right) + \frac{1}{2a^s} + \frac{1}{a^{s-1}} \frac{1}{s-1}$$
$$- \frac{1}{a^{s-1}} \sum_{n=1}^{\infty} \left(e^{-2\pi i n a} \int_0^1 e^{2\pi i n a u} u^{-s} \, du \right.$$
$$\left. + e^{2\pi i n a} \int_0^1 e^{-2\pi i n a u} u^{-s} \, du \right).$$

We invert the order of summation and integration in the last term and consider the series $\sum'^{\infty}_{n=-\infty} e^{-2\pi i n a(u-1)}$ as the Fourier series for $\delta(a(u-1)) - 1$. Then we are left with the integration ($\sigma < 0$)

$$- \int_0^1 \delta(a(u-1)) \, u^{-s} \, du + \int_0^1 u^{-s} \, du = -\frac{1}{2a} - \frac{1}{s-1}.$$

Hence the last term is $-\frac{1}{2a^s} - \frac{1}{s-1} \frac{1}{a^{s-1}}$, which cancels the second term and we finally arrive at the Hurwitz formula

$$\zeta(s,a) = \frac{\Gamma(1-s)}{(2\pi)^{1-s}} \left(e^{\frac{1-s}{2}\pi i} l_{1-s}(1-a) + e^{-\frac{1-s}{2}\pi i} l_{1-s}(a) \right), \qquad (3.67)$$

which is equivalent to (3.48).

Finally, we introduce a class of functions $\gamma_n(x)$, $n \in \mathbb{N}$ (due to Milnor [Mi]) defined by

$$\gamma_{1-t}(x) = \frac{\partial}{\partial t} \zeta(t,x) \qquad (3.68)$$

Note that

$$\gamma_1(x) = \zeta'(0, x) \left(= \log \frac{\Gamma(x)}{\sqrt{2\pi}} \right).$$

Exercise 3.1   Prove the Kubert identity

$$\zeta(1 - s, x) = q^{s-1} \sum_{k=0}^{q-1} \zeta\left(1 - s, \frac{x+k}{q} \right) \tag{3.69}$$

for each $n \in \mathbb{N}$ ($s \neq 0$). Also prove the modified Kubert identity

$$\gamma_n(x) = (\log q) \frac{1}{n} B_n(x) + q^{n-1} \sum_{k=0}^{q-1} \gamma_n\left(\frac{x+h}{q} \right). \tag{3.70}$$

Solution   By (8.12),

$$\Phi(s, a, 1) = \zeta(s, a). \tag{3.71}$$

Hence (8.13) reduces to (3.69).

To prove (3.70), we differentiate

$$q^s \zeta(s, x) = \sum_{k=0}^{q-1} \zeta\left(s, \frac{x+k}{q} \right), \tag{3.72}$$

with respect to s to obtain

$$\zeta'(1 - n, x) = -(\log q) \zeta(1 - n, x) + q^{n-1} \sum_{k=0}^{q-1} \zeta'\left(1 - n, \frac{x+k}{q} \right).$$

This leads to (3.70) on appealing to (4.1).

For more information on Kubert identities, the reader is referred to Sun [Su1].

Chapter 4

The theory of Bernoulli polynomilas via zeta-functions

Abstract

In this chapter we shall deduce some of the basic properties of Bernoulli polynomials from those of the Hurwitz zeta-function. The basis is the relation (4.1). We may develop the theory of Euler polynomials in the same spirit. This is due to the fact that the Euler number E_n corresponds to the special value $L(-n, \chi_4)$ of the Dirichlet L-function $L(s, \chi_4)$ with the unique odd character mod 4 (cf. Chapter 8), which therefore is not presented (cf. e.g. [SC]).

Exercise 4.1   Under (U) deduce

$$\zeta(1 - n, x) = -\frac{1}{n} \overline{B}_n(x), \quad n \in \mathbb{N} \tag{4.1}$$

from (3.8).

Solution   Formula (3.8) with $u = n - 1$, $l = n$ reads for $0 < a < 1$,

$$\zeta(1 - n, a) = a^{n-1} - \frac{1}{n} a^n - \sum_{r=1}^{n} \frac{(-1)^r}{r} \binom{n-1}{r-l} B_r \, a^{n-r}.$$

This can be transformed, on using (cf. (1.13) and (1.16))

$$B_r = (-1)^r B_r, \quad r \geq 2, \quad B_1 = -\frac{1}{2},$$

into

$$\zeta(1-n,a) = -\frac{1}{n}a^n + \frac{1}{2}a^{n-1} - \frac{1}{n}\sum_{r=2}^{n}\binom{n}{r}B_r\, a^{n-r}$$

$$= -\frac{1}{n}\sum_{r=0}^{n}\binom{n}{r}B_r\, a^{n-r},$$

which is (4.1) in view of (1.7).

Remark 4.1 *Since (3.8) depends on (1.1), we have deduces (4.1) under (D') and (U). But since (D'), (A) and (U) are equivalent as giving Taylor coefficients, we may choose any one of them as a definition and assume other two valid.*

We may also define the Bernoulli polynomial $B_s(x)$ as an integral function of s through the relation

$$B_s(x) = -s\,\zeta(1-s,x) \tag{4.2}$$

as in [Ca] or [Mi] and develop the whole theory independently of Chapter 1 (which procedure will be sketched below), and thus we shall take (4.1) for granted and deduce other properties from (3.8) etc.

As a special case of Exercise 4.1, we have ($n \in \mathbb{N}$)

$$\zeta(1-n) = -\frac{1}{n}B_n(1) = \begin{cases} -\frac{1}{n}B_n, & n \geq 2, \\ B_1 = -\frac{1}{2}, & n = 1, \end{cases} \tag{4.3}$$

by (1.16), which in turn is a consequence of (1.9).

Exercise 4.2 Deduce (1.9) from (3.67) and (A) (and (4.1)).

Solution For $0 < x < 1$, (3.67) reads for $s = 1-n$, $n \in \mathbb{N}$

$$\zeta(1-n,x) = \frac{\Gamma(n)}{(2\pi)^n}\{e^{-\frac{\pi i}{2}n}\,l_n(x) + e^{\frac{\pi i}{2}n}\,l_n(1-x)\}, \tag{4.4}$$

so that

$$\overline{B}_n(x) = -n\,\zeta(1-n,x)$$
$$= -\frac{\Gamma(n+1)}{(2\pi)^n}\{(-i)^n\,l_n(x) + i^n\,l_n(1-x)\} \tag{4.5}$$

and

$$\overline{B}_n(1-x) = -\frac{\Gamma(n+1)}{(2\pi)^n}\{(-i)^n\,l_n(1-x) + i^n\,l_n(x)\}.$$

Comparing these completes the proof of (1.9) in the case $0 < x < 1$.

If, in general, $n - 1 < x < n$, $n \in \mathbb{Z}$, then $[x] = n$ and $[1 - x] = -n$, and therefore

$$[1 - x] = -[x], \quad x \notin \mathbb{Z}. \tag{4.6}$$

By (A),

$$B_n(x) = B_n(x + [x]) = \sum_{k=0}^{n} \binom{n}{k} \overline{B}_{n-k}(x)[x]^k.$$

Substituting (1.9) for $0 < x < 1$ and (4.6). we conclude (1.9) for $x \notin \mathbb{Z}$. For $x \in \mathbb{Z}$, (1.9) follows from continuity.

Proposition 4.1 *The difference relation (DE) for Bernoulli polynomials is a consequence of that for the Hurwitz zeta-function (3.4).*

Proof. This follows immediately from (3.4) (under (4.1)):

$$B_n(x + 1) - B_n(x) = -n\{\zeta(1 - n, x + 1) - \zeta(1 - n, x)\}$$
$$= nx^{n-1}.$$

$\square$

Proposition 4.2 *The functional equation (3.67) for the Hurwitz zeta-function implies the Fourier expansion (H) for the Bernoulli polynomials.*

Proof. (4.5) reads

$$\overline{B}_n(x) = -\frac{n!}{(2\pi i)^n} \left\{ \sum_{k=1}^{\infty} \frac{e^{2\pi i k x}}{k^n} + (-1)^n \sum_{k=1}^{\infty} \frac{e^{2\pi i(-k)x}}{k^n} \right\},$$

which gives (H) for $n \geq 2$, since, then, the series are absolutely convergent.

In the case $n = 1$, the sum is to be taken symmetrically:

$$\lim_{N \to \infty} \sum_{\substack{k=-N \\ k \neq 0}}^{N} \left(\frac{e^{2\pi i k x}}{k} - \frac{e^{-2\pi i k x}}{k} \right) = \lim_{N \to \infty} 2i \sum_{k=1}^{N} \frac{\sin 2\pi k x}{k},$$

whence

$$\overline{B}_1(x) = -\frac{1}{\pi} \sum_{k=1}^{\infty} \frac{\sin 2\pi k x}{k}$$

as in (7.9).

$\square$

Proposition 4.3   *The Kubert identity* (3.69) *for the Hurwitz zeta-function implies the Kubert identity* (1.8) *for Bernoulli polynomials.*

Proof.   This follows from (3.69) on substituting (4.1).   □

Now that we have established (DE), (H) and (K) for Bernoulli polynomials, we may trace the logical path given at the end of Chapter 1 to complete the theory of Bernoulli polynomials.

Chapter 5

The theory of the gamma and related functions via zeta-functions

Abstract

In this chapter we shall give a new foundation of the theory of the gamma and related functions. The core of the idea lies in appealing to Lerch's formula (5.4) through which we may transfer the results on the Hurwitz zeta-function to the gamma function (cf. [Mi]), as was the case with the Bernoulli polynomials and the function $\zeta(0, x)$ in Chapter 4.

We shall give two (three if we count the uniqueness theorem as one) proofs of Lerch's formula with minimum possible assumptions: the integral representation (5.1) for $\zeta(s, z)$ (which is a corollary to Theorem 3.1 and the value $\Gamma\left(\frac{1}{2}\right)$). Then we continue to keep the assumption minimum by defining the digamma function by either of the conditions in Lemma 5.1 and the gamma function as its integral.

5.1 Derivatives of the Hurwitz zeta-function

Notation: $s = \sigma + it$ is the complex variable; z is another complex variable, used interchangeably with s;

$$\psi(s) = -\lim_{N \to \infty} \left(\sum_{n=0}^{N} \frac{1}{n+s} - \log(N+s) \right)$$

—the digamma function (cf. (5.18));

$$\log \Gamma(s) = \int_1^s \psi(u)\, du$$

—the log-gamma function;

$$\gamma = -\psi(1)$$

—the Euler's constant (cf. (5.16));

$$\zeta(s, a) = \sum_{n=0}^{\infty} \frac{1}{(n+a)^s}$$

—the Hurwitz zeta-function, where $\sigma > 1$, $a \in \mathbb{C}$, $a \neq$ non-negative integer;

$$\zeta(s) = \zeta(s, 1) = \sum_{n=1}^{\infty} \frac{1}{n^s}, \qquad \sigma > 1$$

—the Riemann zeta-function. For $x \geq 0$, $a \in \mathbb{C}$, $a \neq$ non-negative integer, $u \in \mathbb{C}$,

$$L_u(x, a) = \sum_{0 \leq n \leq x} (n+a)^u$$

—the partial sum ((3.5)) of the Hurwitz zeta-function $\zeta(-u, a)$;

$$\overline{B}_k(t) = B_k(t - [t])$$

—the k-th periodic Bernoulli polynomial;

$$B_k(t) = \sum_{r=0}^{k} \binom{k}{r} B_r\, t^{k-r}$$

—the k-th Bernoulli polynomial (cf. (1.7)); B_k-the k-th Bernoulli number; $[t]$ —the integral part of t.

Theorem 5.1 *If we suppose the integral representations for $\zeta(s, z)$ and $\psi(z)$:*

$$\zeta(s, z) = \frac{1}{s-1} z^{1-s} + \frac{1}{2} z^{-s} - s \int_0^{\infty} \overline{B}_1(t)\, (t+z)^{-s-1}\, dt, \quad \sigma > -1, \quad (5.1)$$

$$\psi(z) = \log z - \frac{1}{2} z^{-1} + \int_0^{\infty} \overline{B}_1(t)\, (t+z)^{-2}\, dt, \qquad (5.2)$$

and also the value

$$\Gamma\left(\frac{1}{2}\right) = \pi^{\frac{1}{2}} \tag{5.3}$$

as known, then we have **Lerch's formula**

$$\zeta'(0, z) = \log \frac{\Gamma(z)}{\sqrt{2\pi}}. \tag{5.4}$$

Proof. Integrating (5.2) from 1 to z, we obtain

$$\log \Gamma(z) = \int_1^z \left(\log z - \frac{1}{2z}\right) dz + \int_0^\infty \overline{B}_1(t) \int_1^z (t + z)^{-2} \, dz \, dt$$

$$= z \log z - z - \frac{1}{2} \log z + 1 \tag{5.5}$$

$$- \int_0^\infty \overline{B}_1(t) \, (t + z)^{-1} \, dt + \int_1^\infty \overline{B}_1(t) \, t^{-1} \, dt.$$

On the other hand, differentiation of (5.1) gives

$$\zeta'(0, z) = z \log z - z - \frac{1}{2} \log z - \int_0^\infty \overline{B}_1(t) \, (t + z)^{-1} \, dt. \tag{5.6}$$

Comparing (5.5) and (5.6), we see that

$$\zeta'(0, z) = \log \Gamma(z) - 1 - \int_1^\infty \overline{B}_1(t) \, t^{-1} \, dt \tag{5.7}$$

and it remains to evaluate the last integral.

For this we differentiate the formula $\zeta\left(s, \frac{1}{2}\right) = (2^s - 1)\zeta(s)$ to obtain

$$\zeta'\left(s, \frac{1}{2}\right) = 2^s (\log 2) \zeta(s) + (2^s - 1) \zeta'(s). \tag{5.8}$$

Hence in view of $\zeta(0) = -\frac{1}{2}$, a consequence of (5.1),

$$\zeta'\left(0, \frac{1}{2}\right) = (\log 2) \zeta(0) = -\frac{1}{2} \log 2. \tag{5.9}$$

Now put $z = \frac{1}{2}$ in (5.7) and use the value of $\Gamma(\frac{1}{2})$ to obtain

$$-\frac{1}{2} \log 2 = \log \sqrt{\pi} - 1 - \int_1^\infty \overline{B}_1(t) \, t^{-1} \, dt. \tag{5.10}$$

Hence

$$1 + \int_1^\infty \overline{B}_1(t) \, t^{-1} \, dt = \log \sqrt{2\pi}, \tag{5.11}$$

which is of some interest in its own right.

(5.6) and (5.11) combine to give (5.4). This completes the proof. □

Exercise 5.1   If we assume the integral representation (5.1) and the value

$$\zeta'(0) = -\log\sqrt{2\pi} \tag{5.12}$$

as known, then we have Lerch's formula (5.4).

Proof.   Differentiation of (5.11) with respect to s gives rise to

$$\zeta'(s,x)$$

$$= \frac{-1}{(s-1)^2} x^{1-s} - \frac{x^{1-s}}{s-1}\log x - \frac{1}{2}x^{-s}\log x - \int_0^\infty \overline{B}_1(t)\,(t+x)^{-s-1}\,\mathrm{d}t$$

$$- s\int_0^\infty \overline{B}_1(t)\,(t+x)^{-s-1}\,(-\log(t+x))\,\mathrm{d}t,$$

whence

$$\zeta'(0,x) = -x + x\log x - \frac{1}{2}\log x - \int_0^\infty \overline{B}_1(t)\,(t+x)^{-1}\,\mathrm{d}t. \tag{5.13}$$

Now

$$\frac{\partial^2}{\partial x^2}\zeta'(0,x) = \frac{1}{x} + \frac{1}{2}\frac{1}{x^2} - 2\int_0^\infty \overline{B}_1(t)\,(t+x)^{-3}\,\mathrm{d}t. \tag{5.14}$$

Now the last integral on the right-hand side of (5.14) is the sum of the terms

$$\int_n^{n+1} \left(t - n - \tfrac{1}{2}\right)(t+x)^{-3}\,\mathrm{d}t$$

$$= \int_n^{n+1} \left((t+x) - \left(n+x+\tfrac{1}{2}\right)\right)(t+x)^{-3}\,\mathrm{d}t$$

$$= \int_n^{n+1} \left((t+x)^{-2} - \left(n+x+\tfrac{1}{2}\right)(t+x)^{-3}\right)\,\mathrm{d}t$$

$$= \left[-\frac{1}{t+x} + \frac{1}{2}\left(n+x+\tfrac{1}{2}\right)(t+x)^{-2}\right]_n^{n+1}$$

$$= \frac{1}{n+x} - \frac{1}{n+x+1}$$

$$\quad + \frac{1}{2}\left(n+1+x-\tfrac{1}{2}\right)(n+1+x)^{-2} - \frac{1}{2}\left(n+x+\tfrac{1}{2}\right)(n+x)^{-2}$$

$$= \frac{1}{n+x} - \frac{1}{n+x+1}$$

$$+ \frac{1}{2} \left(\frac{1}{n+1+x} - \frac{1}{n+x} \right) - \frac{1}{4} \left(\frac{1}{(n+1+x)^2} + \frac{1}{(n+x)^2} \right).$$

Hence

$$\int_0^\infty \overline{B}_1(t)(t+x)^{-3} \, \mathrm{d}t$$

$$= \frac{1}{2} \sum_{n=0}^\infty \left(\frac{1}{n+x} - \frac{1}{n+x+1} \right) - \frac{1}{2} \sum_{n=0}^\infty \frac{1}{(n+x)^2} + \frac{1}{4} \frac{1}{x^2}$$

$$= \frac{1}{2} \frac{1}{x} + \frac{1}{4} \frac{1}{x^2} - \frac{1}{2} \sum_{n=0}^\infty \frac{1}{(n+x)^2}.$$

Substituting this in (5.14), we obtain

$$\frac{\partial^2}{\partial x^2} \zeta'(0,x) = \frac{1}{x} + \frac{1}{2} \frac{1}{x^2} - \frac{1}{x} - \frac{1}{2} \frac{1}{x^2} + \sum_{n=0}^\infty \frac{1}{(n+x)^2}$$

$$= \zeta(2,x).$$

It is essential to notice that

$$\frac{\mathrm{d}^2}{\mathrm{d}x^2} \log \Gamma(x) = \frac{\mathrm{d}}{\mathrm{d}x} \frac{\Gamma'}{\Gamma}(x) = \frac{\mathrm{d}}{\mathrm{d}x} \psi(x) = \sum_{n=0}^\infty \frac{1}{(n+x)^2},$$

the last being due to (5.17) below.

Hence

$$\frac{\partial^2}{\partial x^2} \zeta'(0,x) = \frac{\mathrm{d}^2}{\mathrm{d}x^2} \log \Gamma(x). \tag{5.15}$$

(5.15) gives rise to

$$\zeta'(0,x) = \log \Gamma(x) + ax + b.$$

First

$$\zeta'(0,1) = a + b$$

and

$$\zeta'(0,2) = 2a + b.$$

Recalling $\zeta(s, x+1) = \zeta(s,x) - x^{-s}$, we see that $\zeta'(s, x+1) = \zeta'(s,x) + x^{-s} \log x$, whence $\zeta'(0,2) = \zeta'(0,1)$. Hence $a = 0$.

The value of $b = \zeta'(0,1) = \zeta'(0)$ is determined by (5.12) and we have

$$\zeta'(0,x) = \log \Gamma(x) - \log \sqrt{2\pi},$$

i.e. Lerch's formula. □

Lemma 5.1 *Under the definition of* **Euler's constant**

$$\gamma = \lim_{N \to \infty} \left(\sum_{n=1}^{N} \frac{1}{n} - \log(N + z) \right) \tag{5.16}$$

for any z other than negative integers, the two definitions for ψ are equivalent:

$$\psi(z) + \gamma = \sum_{n=1}^{\infty} \left(\frac{1}{n} - \frac{1}{z + n - 1} \right), \tag{5.17}$$

the **Gaussian representation** *(cf. (5.41)), and*

$$\psi(z) = - \lim_{N \to \infty} \left(\sum_{n=0}^{N} \frac{1}{n + z} - \log(N + z) \right), \tag{5.18}$$

for any z other than negative integers, the **generic definition**.

Proof. Substituting (5.16) in (5.17) in the form

$$\psi(z) + \gamma = \lim_{N \to \infty} \sum_{n=1}^{N} \left(\frac{1}{n} - \frac{1}{z + n + 1} \right)$$

$$= \lim_{N \to \infty} \left(\sum_{n=1}^{N} \left(\frac{1}{n} - \log(N + z) \right) \right.$$

$$\left. - \sum_{n=1}^{N} \left(\frac{1}{z + n - 1} - \log(N + z) \right) \right),$$

we deduce that

$$\psi(z) + \gamma = \gamma - \lim_{N \to \infty} \sum_{n=0}^{N} \left(\frac{1}{z + n} - \log(N + z) \right),$$

whence (5.18).

On the other hand, (5.18) may be written as

$$\psi(z) = \lim_{N \to \infty} \left(\sum_{n=1}^{N} \left(\frac{1}{n} - \frac{1}{z + n - 1} \right) - \sum_{n=1}^{N} \left(\frac{1}{n} - \log(N + z) \right) \right),$$

$$= \sum_{n=1}^{\infty} \left(\frac{1}{n} - \frac{1}{z + n - 1} \right) - \gamma,$$

i.e. (5.17). □

Remark 5.1 *In (5.16), z is usually taken to be 0, but can be any number as in (5.16) because*

$$\log(N + z) - \log(N + w) \to 0, \quad N \to \infty.$$

The absolute convergence of the series in (5.17) is clear because each term is $O\left(\frac{1}{n^2}\right)$, and the existence of limits in (5.16) and (5.18) follows from the comparison with the corresponding integral, or the Euler-Maclaurin formula (cf. Chapter 2).

We shall illustrate the far-reaching power of Theorem 3.1 by the first derivative $\left(\frac{\partial}{\partial u} L_u(x, a)\right.$ or $\left.-\zeta'(-u, a)\right)$ in the special case of $u = m$, $m \in \mathbb{N} \cup \{0\}$. For $\mathbb{N} \ni l > m + 1$, Corollary 3.1 eventually yields (cf. [KTTY3])

$$-\zeta'(-m, a) = \lim_{N \to \infty} \left(\sum_{n=0}^{N} (n + a)^m \log(n + a) \right. \tag{5.19}$$

$$- \frac{1}{m + 1}(N + a)^{m+1} \log(N + a) + \frac{1}{(m + 1)^2}(N + a)^{m+1}$$

$$- \frac{1}{2}(N + a)^m \log(N + a) - \sum_{r=2}^{m+1} \binom{m}{r - 1} \frac{B_r}{r!}$$

$$\left. \cdot \left(\frac{1}{m} + \cdots + \frac{1}{m - r + 2} + \log(N + a) \right) (N + a)^{m-r+1} \right).$$

and

$$\zeta'(-m, a) \tag{5.20}$$

$$= \frac{1}{m + 1} a^{m+1} \log a - \frac{1}{(m + 1)^2} a^{m+1} - \frac{1}{2} a^m \log a + \frac{1}{12} a^{m-1} \log a$$

$$+ \sum_{r=4}^{m+1} \frac{B_r}{r} \left(\sum_{j=0}^{r-2} (-1)^j \binom{m}{j} \frac{1}{r - 1 - j} + \binom{m}{r - 1} \log a \right)$$

$$+ \frac{1}{m + 1} \sum_{r=m+2}^{l} B_r \left(\sum_{j=0}^{r-1} (-1)^j \binom{r - m - 2}{j} \frac{1}{r - j} \right) a^{m-r+1}$$

$$+ (-1)^{l+1} \int_0^\infty \left(\sum_{j=0}^{l-1} (-1)^j \binom{l - m - 1}{j} \frac{1}{l - j} \overline{B}_l(t)(t + a)^{m-l} \right) dt,$$

where (5.19) and (5.20) correspond to (3.7) and (3.8), respectively.

Exercise 5.2   Show that, in view of Lerch's formula (5.4), (5.19) with $m = 0$ gives Euler's product formula (or Weierstrass' canonical product of genus 1) (5.21) for $\Gamma(a)$.

$$\frac{1}{\Gamma(a)} = a \prod_{n=1}^{\infty} \left\{ \left(1 + \frac{1}{n}\right)^{-a} \left(1 + \frac{a}{n}\right) \right\}, \tag{5.21}$$

Solution   Indeed, (5.19) with $m = 0$ reads

$$-\zeta'(0, a) = \lim_{N \to \infty} \left(\sum_{n=0}^{N} \log(n + a) - \frac{1}{2} \log(N + a) \right. \tag{5.22}$$
$$\left. - (N + a) \log(N + a) + N + a \right).$$

For $a = 1$, (5.22) with $N + 1$ replaced by N gives

$$-\zeta'(0) = \lim_{N \to \infty} \left(\sum_{n=1}^{N} \log n - \left(N + \frac{1}{2}\right) \log N + N \right). \tag{5.23}$$

Substituting

$$\frac{1}{2} \log(N + a) = \frac{1}{2} \log N + o(1),$$
$$(N + a) \log(N + a) = N \log N + a \log N + a + o(1),$$

we transform (5.22), under (5.4), into

$$-\log \frac{\Gamma(a)}{\sqrt{2\pi}}$$
$$= \lim_{N \to \infty} \left(\sum_{n=0}^{N} \log(n + a) - \left(N + \frac{1}{2}\right) \log N + N - a \log N \right). \tag{5.24}$$

Subtracting (5.23) from (5.24) yields, on using (2.20),

$$-\log \Gamma(a) = \lim_{N \to \infty} \left(\log a + \sum_{n=1}^{N} \log \frac{n + a}{n} - a \log N \right). \tag{5.25}$$

By expressing $\log N$ as $\sum_{n=1}^{N-1} \log \frac{n+1}{n}$, we may write

$$
-\log \Gamma(a) = \lim_{N \to \infty} \left(\log a + \sum_{n=1}^{N} \log \left\{ \frac{n+a}{n} \cdot \left(\frac{n+1}{n} \right)^{-a} \right\} \right) \tag{5.26}
$$

$$
= \log a + \sum_{n=1}^{\infty} \log \left\{ \left(1 + \frac{1}{n} \right)^{-a} \left(1 + \frac{a}{n} \right) \right\},
$$

whence (5.21).

Remark 5.2   *Our procedure is a reverse to that of Berndt [Ber2] in which he starts from one of the equivalent definitions of the gamma function given by*

$$
\Gamma(a) = \lim_{N \to \infty} \frac{N! \, (N+1)^a}{a(a+1) \cdots (a+N)}, \tag{5.27}
$$

Euler's interpolation formula or

$$
\log \Gamma(a) = \lim_{N \to \infty} \left(-\sum_{n=0}^{N} \log(n+a) + \sum_{n=1}^{N} \log n + a \log(N+1) \right) \tag{5.27$'$}
$$

and deduces Lerch's formula by comparing (5.27)$'$ with (5.22) (and (5.23)).

Of course, we can cover (5.27)$'$ in the same way as above. Indeed, from (5.23) and (5.24), we deduce that

$$
-\zeta'(0,a) + \zeta'(0) = \lim_{N \to \infty} \left(\sum_{n=0}^{N} \log(n+a) - \sum_{n=1}^{N} \log n - a \log N \right),
$$

which reduces to (5.27)$'$ by Lerch's formula, save for the value $\zeta'(0) = -\log \sqrt{2\pi}$; this value is found in Exercise 2.6 with the Stirling's formula being taken for granted.

We now turn to recover Deninger's Theorem 2.3 [D], especially, the Gaussian representation

$$
-\zeta''(0,a) = -\zeta''(0) - \log^2 a
$$

$$
+ \lim_{N \to \infty} \left(a \log^2 N - \sum_{n=1}^{N} \left(\log^2(n+a) - \log^2 n \right) \right). \tag{5.28}
$$

Indeed, choosing $u = 0$ and $x = N \in \mathbb{N}$ in Corollary 3.2, we get

$$\sum_{n=0}^{N} \log^2(n+a) = \frac{1}{2} \log^2(N+a) - \int_{N}^{\infty} \frac{\overline{B}_1(t)}{t+a} \log^2(t+a)\, dt \qquad (5.29)$$
$$+ (N+a)\left\{\log^2(N+a) - 2\log(N+a) + 2\right\} + \zeta''(0, a).$$

Put $a = 1$ and write N for $N+1$ in (5.29) to get

$$\sum_{n=1}^{N} \log^2 n = \frac{1}{2} \log^2 N - \int_{N}^{\infty} \frac{\overline{B}_1(t)}{t} \log^2 t\, dt \qquad (5.30)$$
$$+ N\left(\log^2 N - 2\log N + 2\right) + \zeta''(0).$$

Noting that $\log(N+a) = \log N + \frac{a}{N} + O\left(\frac{1}{N^2}\right)$, we have $\log^2(N+a) = \log^2 N + \frac{2a}{N} \log N + O\left(\frac{1}{N^2}\right)$. Hence

$$(N+a)\{\log^2(N+a) - 2\log(N+a) + 2\}$$
$$= N\log^2 N - 2N\log N + 2N + a\log^2 N + O\left(\frac{1}{N}\right).$$

Hence (5.29) may be written as

$$\sum_{n=1}^{N} \log^2(n+a) = \frac{1}{2}\log^2 N - \int_{N}^{\infty} + N\log^2 N - 2N\log N + 2N \qquad (5.29)'$$
$$+ a\log^2 N - \log^2 a + \zeta''(0, a) + O\left(\frac{1}{N}\right).$$

From (5.29)' and (5.30) it follows that

$$-\zeta''(0, a) = -\zeta''(0) + a\log^2 N - \log^2 a - \sum_{n=1}^{N}\left(\log^2(n+a) - \log^2 n\right)$$
$$- \int_{N}^{\infty} \overline{B}_1(t)\left(\frac{\log^2(t+a)}{t+a} - \frac{\log^2 t}{t}\right) dt + O\left(\frac{1}{N}\right),$$

or

$$-\zeta''(0, a) = -\zeta''(0) - \log^2 a + a\log^2 N$$
$$- \sum_{n=1}^{N}\left(\log^2(n+a) - \log^2 n\right) + O\left(\frac{\log^2 N}{N}\right) \qquad (5.31)$$

upon estimating the integral. Now (5.31) implies (5.28).

We may also recover the Weierstrass representation ([D, (2.3.2)]) by Corollary 3.1 with $u = -1$, $a = 1$ and $x = N \in \mathbb{N}$ (we write N for $N + 1$):

$$\sum_{n=1}^{N} \frac{\log n}{n} = \frac{1}{2} N^{-1} \log N - \int_{N}^{\infty} \frac{\overline{B}_1(t)}{t^2} \log t \, dt + \frac{1}{2} \log^2 N + \gamma_1$$

or

$$\gamma_1 = \sum_{n=1}^{N} \frac{\log n}{n} - \frac{1}{2} \log^2 N + O\left(\frac{\log N}{N}\right). \tag{5.32}$$

(cf. [KKSY, (8)]).

Solving (5.32) for $\log^2 N$ and substituting it in (5.31), we deduce that

$$-\zeta''(0, a) = -\zeta''(0) - 2\gamma_1 \log a - \log^2 a \tag{5.33}$$

$$- \sum_{n=1}^{N} \left(\log^2(n + a) - \log^2 n - 2\frac{\log n}{n}\right) + O\left(\frac{\log^2 N}{N}\right).$$

5.2 Asymptotic formulas for the Hurwitz and related zeta-functions in the second variable

In this section we shall show that our formula $(3.8)'$ below coincides with Katsurada's formula (2.2) (Theorem 1 of [Kat1]) in the special case when $\lambda = 1$. Since our formula (3.44) with confluent hypergeometric function coefficients readily extends to a general l, it suffices to show that the main terms coincide with each other.

We suppose that $u \neq -1$ and apply Formula (3.8) with $\alpha + z$ in place of a, and the relations $\binom{u}{r-1}\frac{1}{r} = \frac{1}{u+1}\binom{u+1}{r}$ and

$$(-1)^k B_k = B_k(1) = \begin{cases} B_k, & k \neq 1 \\ B_1 + 1, & k = 1, \end{cases}$$

(cf. (1.16)) in order to obtain

$$\zeta(-u, \alpha+z) = -\frac{1}{u+1} \sum_{r=0}^{l} \binom{u+1}{r} (\alpha+z)^{u-r+1} B_r + O\left(|z|^{\operatorname{Re}(u)-l}\right). \tag{3.8}'$$

Supposing further that $|\alpha| < |z|$, we infer, by the binomial expansion, that the right-hand side of $(3.8)'$ can be written as

$$S + O\left(|z|^{\operatorname{Re}(u)-l}\right),$$

where

$$S := -\frac{1}{u+1} \sum_{r=0}^{l} \sum_{k=r}^{l} \binom{u+1}{r} \binom{u+1-r}{k-r} B_r \, \alpha^{k-r} z^{u-k+1},$$

since, for $k \geq l - r$, we have Re $(u) - r - k \leq$ Re $(u) - l$.

Using (1.14), we obtain

$$S = -\frac{1}{u+1} \sum_{k=0}^{l} \binom{u+1}{k} z^{u-k+1} z^{u-k+1} \sum_{r=0}^{k} \binom{k}{r} B_r \, \alpha^{k-r},$$

whose innermost sum is precisely $B_k(\alpha)$.

Hence we conclude that

$$S = -\frac{1}{u+1} z^{u+1} + \sum_{r=0}^{l-1} \frac{(-1)^{r+1}}{(r+1)!} (-u)_r \, B_{r+1}(\alpha) \, z^{u-r}. \qquad (5.34)$$

Substituting (5.34) into (3.8)′, we obtain the special case of Katsurada's result [Kat1, p.168, Theorem 1] when $\lambda = 1$.

Theorem 5.2   *For any integer $l \geq 0$ and any z in $|\arg(z)| < \pi$,*

$$\zeta(s, \alpha + z) = \frac{1}{s-1} z^{1-s} + \sum_{r=0}^{l-1} \frac{(-1)^{r+1}}{(r+1)!} B_{r+1}(\alpha) \, (s)_r \, z^{-s-r}$$
$$+ O\left(|z|^{-\operatorname{Re}(s)-l}\right). \qquad (5.35)$$

Remark 5.3   *Formula (5.35), in conjunction with a generalization of Formula (3.64) will yield the aforementioned special case of Katsurada's main result [Kat1, p.168, Formula (2.2)].*

The method of proof of Theorem 5.2 readily extends to $-\zeta'(s, \alpha)$ and $\zeta''(s, \alpha)$ contained in Corollary 3.1 and Corollary 3.2, respectively. Thus we have the following consequences of Theorem 3.1 corresponding to Katsurada's Corollary 1 and Corollary 2 in [Kat1].

We restate these results (see Corollary 3.3 and Corollary 3.2) in terms of $\log \Gamma(z)$ and $R(z)$ in view of Lerch's formula (5.4) and Deninger's definition:

$$R(z) = -\zeta''(0, z). \qquad (5.36)$$

Corollary 5.1   *For any integer l and any fixed $\alpha > 0$, the generalized*

Stirling's formula:

$$\log\left(\frac{\Gamma(z+\alpha)}{\sqrt{2\pi}}\right) = \left(z+\alpha-\frac{1}{2}\right)\log z - z$$
$$+ \sum_{r=1}^{l-1} \frac{(-1)^{r+1}}{r(r+1)} B_{r+1}(\alpha)\, z^{-r} + O\left(|z|^{-l}\right)$$

holds true for $z \to \infty$ *and* $|\arg(z)| \le \pi - \delta$ $(\delta > 0)$.

Corollary 5.2 *For any integer* $l \ge 1$ *and fixed* $\alpha > 0$,

$$R(z+\alpha) = \left(z+\alpha-\frac{1}{2}\right)(\log z)^2 - 2z\log z + 2z$$
$$+ \sum_{r=1}^{l-1} \frac{(-1)^{r+1}}{r(r+1)} B_{r+1}(\alpha)\, z^{-r}\left(\log z - \sum_{h=1}^{r-1}\frac{1}{h}\right) \qquad (5.37)$$
$$+ O\left(|z|^{-l}\log(|z|+1)\right)$$

valid for $z \longrightarrow \infty$ *and* $|\arg(z)| \le \pi - \delta$ $(\delta > 0)$

5.3 An application of the Euler digamma function

In this section what we are going to mainly use is the case $u = -1$ of Theorem 3.1, which we restate as the following:

Theorem 5.3 *For* $x \ge 0$, $a \in \mathbb{C}$, $a \ne$*non-negative integer, we have*

$$L_{-1}(x,a) = \log(x+a) - \psi(a)$$
$$- \sum_{r=1}^{l} \frac{1}{r}\overline{B}_r(x)\,(x+a)^{-r} + \int_x^\infty \overline{B}_l(t)\,(t+a)^{-1-l}\,dt. \qquad (5.38)$$

Corollary 5.3 *(i)* (3.14) *is a special case of* (5.38) *with* $x = 0$.
(ii) (5.18) *is a special case of* (5.38) *as* $x \to \infty$.
(iii) $\psi(a)$ *admits the Gaussian representation*

$$\psi(a) = \lim_{N\to\infty}\left(-z + \frac{a}{N} - \frac{1}{a} - \sum_{k=1}^{N}\left(\frac{1}{k+a} - \frac{1}{k}\right)\right). \qquad (5.39)$$

Proof. Only (5.39) needs a proof, which goes on the similar lines as those for the case $u = -1$ of Theorem 3.1. Formula (5.38) gives

$$\psi(a) = \log(x + a) - \sum_{0 \le n \le x} \frac{1}{n + a} + O\left(\frac{1}{x}\right),$$

which we slightly rewrite as

$$\psi(a) = \log \frac{N + a}{N} - \sum_{n=1}^{N} \left(\frac{1}{n + a} - \frac{1}{n}\right) + \log N - \sum_{n=1}^{N} \frac{1}{n} + O\left(\frac{1}{N}\right).$$

Recalling (5.16) (which fact is also contained in Theorem 5.3), we deduce (5.39), on taking the limit as $N \to \infty$. $\qquad \square$

Now we shall see what formula (5.39) means in the light of the Dufresnoy-Pisot type uniqueness theorem (cf. [D]).

Lemma 5.2 *If the function* $g : \mathbb{R}_+ \to \mathbb{R}$ *($\mathbb{R}_+$ meaning positive reals) satisfies*

$$\lim_{n \to \infty} (g(x + n) - g(n)) = 0, \quad 0 < x \le 1, \tag{5.40}$$

then for any $\lambda \in \mathbb{R}$ *there is at most one function* $f : \mathbb{R}_+ \to \mathbb{R}$ *with the following properties:*

(a) $f(1) = \lambda$
(b) f *is convex on some interval* (A, ∞), $A > 0$
(c) f *is a solution of the difference equation* (DE)

$$f(x + 1) - f(x) = g(x), \quad x \in \mathbb{R}_+.$$

If such a function exists, it is given by the Gaussian representation (cf. (5.17))

$$f(x) = \lim_{n \to \infty} \left(\lambda + x\, g(n) - g(x) - \sum_{k=1}^{n-1} (g(x + k) - g(k))\right). \tag{5.41}$$

Theorem 5.4 *(i) The digamma function* $\psi(a)$ *defined by (5.18) is a unique solution (convex for large argument) of the DE*

$$f(x + 1) - f(x) = \frac{1}{x}, \quad x \in \mathbb{R}_+. \tag{5.42}$$

(ii) (5.39) is exactly (5.41), furnished by the Dufresnoy-Pisot type theorem, which already entails Assertion (i).

Proof. For curiosity, we prove (i) without assuming (5.39). With $g(x) = \frac{1}{x}$, (5.40) is satisfied. (a) follows from the definition and (c) follows from (5.18); only (b) remains.

We differentiate (5.2) to obtain

$$\psi''(a) = -\frac{1}{a^3} - \frac{1}{a^2} + 6 \int_0^\infty \overline{B}_1(t)\,(t+a)^{-4}\,dt. \tag{5.43}$$

To express the last integral in closed form is an easy exercise. Indeed, it is the sum of integrals of type $\int_n^{n+1}$. Since

$$\int_n^{n+1} \overline{B}_1(t)(t+a)^{-4}\,dt = \int_n^{n+1} \left(t + a - \left(a + n + \frac{1}{2} \right) \right)(t+a)^{-4}\,dt$$

$$= \int_n^{n+1} (t+a)^{-3}\,dt - \left(a + n + \frac{1}{2} \right) \int_n^{n+1} (t+a)^{-4}\,dt$$

$$= -\frac{1}{2} \left((n+1+a)^{-2} - (n+a)^{-2} \right)$$

$$+ \frac{1}{3} \left(n + a + \frac{1}{2} \right) \left((n+1+a)^{-3} - (n+a)^{-3} \right)$$

$$= -\frac{1}{2} \left((n+1+a)^{-2} - (n+a)^{-2} \right) + \frac{1}{3} \left((n+a+1)^{-2} - \frac{1}{2}(n+1+a)^{-3} \right)$$

$$- \frac{1}{3} \left((n+a)^{-2} + \frac{1}{2}(n+a)^{-3} \right)$$

$$= -\frac{1}{6} \left((n+1+a)^{-2} - (n+a)^{-2} \right) - \frac{1}{6} \left((n+a+1)^{-3} + (n+a)^{-3} \right),$$

summing these for $n = 0, 1, 2, 3, 4, \cdots$, we obtain

$$\int_0^\infty \overline{B}_1(t)\,(t+a)^{-4}\,dt = \frac{1}{6}\,a^{-2} - \frac{1}{6}\,a^{-3} - \frac{1}{6} \sum_{n=0}^\infty \frac{1}{(n+1+a)^3}.$$

Hence

$$\psi''(a) = -\frac{2}{a^3} - \zeta(3, a+1) < 0, \tag{5.44}$$

and (b) follows, whence uniqueness follows from Lemma 5.2. $\qquad\square$

Corollary 5.4 (i) *For $|z| < 1$ we have*

$$\sum_{k=2}^\infty \zeta(k,a)\,z^{k-1} = -\psi(a-z) + \psi(a). \tag{5.45}$$

(*ii*) $\log \frac{\Gamma(z)}{\sqrt{2\pi}}$ *is the unique solution (convex for large argument) of the DE*

$$f(x+1) - f(x) = \log x. \tag{5.46}$$

Proof. (*i*) is well-known and best viewed as the Taylor expansion of $\zeta(s, a)$ in the second variable a (cf. [Klu], [KKaY], [SC]) and the proof is immediate as follows.

The left side of (5.45) is

$$\sum_{k=2}^{\infty} z^{k-1} \sum_{n=0}^{\infty} \frac{1}{(n+a)^k} = \frac{1}{z} \sum_{n=0}^{\infty} \sum_{k=2}^{\infty} \left(\frac{z}{n+a} \right)^k$$

$$= \frac{1}{z} \sum_{n=0}^{\infty} \frac{\left(\frac{z}{n+a} \right)^2}{1 - \frac{z}{n+a}} = \sum_{n=0}^{\infty} \frac{1}{(n+a-z)(n+a)}$$

$$= \lim_{x \to \infty} \left(\sum_{0 \le n \le x} \frac{1}{n+a-z} - \log(x+a-z) + \log(x+a) - \frac{1}{n+a} \right),$$

which is the right side of (5.45), in view of (5.18). □

Assertion (*ii*) is known as the Bohr-Mollerup theorem, and is a consequence of Lemma 5.2. We could cover (*ii*) also by our Theorem 5.3, (*ii*) if only we assume we know the value (5.12) $\zeta'(0) = \log \frac{1}{\sqrt{2\pi}}$. We may also regard (*ii*) as Lerch's formula (5.4) ([Ber2]).

Corollary 5.5   *We have the duplication formula*

$$\psi(2z) = \frac{1}{2}\psi(z) + \frac{1}{2}\psi\left(z + \frac{1}{2}\right) + \log 2, \tag{5.47}$$

and, a fortiori,

$$\Gamma(2z) = 2^{2z-1}\pi^{-\frac{1}{2}}\Gamma(z)\Gamma\left(z + \frac{1}{2}\right). \tag{5.48}$$

Proof. Indeed, using (5.18) in the form

$$
\frac{1}{2}\psi(z) + \frac{1}{2}\psi\left(z + \frac{1}{2}\right)
$$

$$
= \lim_{x\to\infty}\left(\frac{1}{2}\log(x+z)\left(x+z+\frac{1}{2}\right)\right.
$$

$$
\left. - \sum_{0\le 2n\le 2x}\frac{1}{2n+2z} - \sum_{0\le 2n+1\le 2x+1}\frac{1}{2n+1+2z}\right)
$$

$$
= \lim_{x\to\infty}\left(\frac{1}{2}\log 4\left(x+z+\frac{1}{2}\right)^2 - \log 2 - \sum_{0\le n\le 2x+1}\frac{1}{n+2z}\right),
$$

which is $\psi(z) - \log 2$. (5.48) follow from (5.47) if we use $\Gamma\left(\frac{1}{2}\right) = \sqrt{\pi}$. $\square$

Remark 5.4 *The property in Corollary 5.3 is a special case of the Kubert identity (or distribution property) shared by a wide class of functions (cf. (8.13), [Mi], [Su1]).*

5.4 The first circle

Proposition 5.1 *The product representation for the gamma function*

$$
\Gamma(z+1) = e^{-\gamma z}\prod_{n=1}^{\infty} e^{\frac{z}{n}}\left(1 + \frac{z}{n}\right)^{-1} \tag{5.49}
$$

is a consequence of (5.45).

Exercise 5.3 Deduce (5.49) from (5.45).

Solution Integrating (5.45) from 0 to z, we obtain

$$
\sum_{k=2}^{\infty}\frac{\zeta(k,a)}{k}z^k = \log\Gamma(a-z) - \log\Gamma(a) + \psi(a)z, \tag{5.50}
$$

which is also a well-known formula (cf. e.g. [SC]). We need only the special case of (5.50) with $a = 1$, z replaced by $-z$:

$$
\sum_{k=2}^{\infty}\frac{(-z)^k}{k}\zeta(k) = \log\Gamma(z+1) + \gamma z. \tag{5.51}
$$

We now apply a procedure similar to that of the proof of Corollary 5.4. We see that the left side of (5.51) becomes

$$\sum_{n=1}^{\infty} \sum_{k=2}^{\infty} \frac{1}{k} \left(-\frac{z}{n}\right)^k,$$

the inner sum of which can be summed by the elementary formula ($|r| < 1$)

$$\sum_{k=2}^{\infty} \frac{1}{k} r^k = -r - \log(1 - r).$$

Hence

$$\sum_{k=2}^{\infty} \frac{(-z)^k}{k} \zeta(k) = \sum_{n=1}^{\infty} \left(\frac{z}{n} - \log\left(1 + \frac{z}{n}\right)\right). \tag{5.52}$$

Combining (5.50) and (5.51) completes the proof of (5.49).

Proposition 5.2   *The reciprocal relation*

$$\Gamma(s)\,\Gamma(1 - s) = \frac{\pi}{\sin \pi s}, \tag{5.53}$$

for the gamma function is a consequence of the asymmetric form of the functional equation

$$\zeta(1 - s) = 2^{1-s}\pi^{-s}\,\Gamma(s) \cos\left(\frac{\pi s}{2}\right) \zeta(s), \tag{5.54}$$

for the Riemann zeta-function.

Proof.   Changing s by $1 - s$ in (5.54), we deduce its counterpart

$$\zeta(s) = 2^s \pi^{s-1}\,\Gamma(1 - s) \sin\left(\frac{\pi s}{2}\right) \zeta(1 - s). \tag{5.55}$$

Multiplying (5.54) and (5.55) and canceling the common factor $\zeta(s)\,\zeta(1 - s)$, we arrive at (5.53).   □

Remark 5.5   *The proof of Proposition 5.2 is modeled on Eisenstein's 1849 proof (cf. [We]) of the functional equation for the Hurwitz-Lerch zeta-function. A standard proof (cf. e.g. [Leb] and Exercise 2.3) is via the beta function. Use is made of the integral formula $\int_0^\infty \frac{x^{1-z}}{1+x}\,dx = \frac{\pi}{\sin \pi z}$, $0 < \mathrm{Re}\, z < 1$.*

Lemma 5.3   *The asymmetric form (5.54) of the functional equation for the Riemann zeta-function is a consequence of the functional equation for the Hurwitz zeta-function (or the Hurwitz formula)* $(0 < a < 1)$

$$\zeta(s, a) = -i \, (2\pi)^{s-1} \, \Gamma(1-s) \left(e^{\frac{\pi}{2} i s} l_{1-s}(a) - e^{-\frac{\pi}{2} i s} l_{1-s}(1-a) \right), \quad (5.56)$$

where $l_s(a) = \sum_{n=1}^{\infty} \dfrac{e^{2\pi i n a}}{n^s}$ *stands for the polylogarithm function (3.3), Formula (5.56) (which already appeared as (3.48) and (3.67)) in the long run, is a consequence of (5.1).*

A recent proof of (5.56) based on the Fourier expansion of the Dirac delta function can be found in [BKT] or [KTTY3] and is sketched in §3.5. A more laborious but easier proof can be found in [R] (for the Riemann zeta) and [PP] (for the general case). It amounts to completing the integral in

$$
\begin{aligned}
\zeta(s, a) = {} & \frac{1}{s-1} \, a^{1-s} + \frac{1}{2} \, a^{-s} + \frac{1}{12} \, s \, a^{-s-1} \\
& - \frac{s(s+1)}{2} \int_0^{\infty} \overline{B}_2(t) \, (t+a)^{-s-2} \, dt, \quad \sigma > -2,
\end{aligned}
\quad (5.57)
$$

in the from $\int_{-a}^{\infty} \overline{B}_2(t)(t+a)^{-s-2} \, dt$, then using the absolutely converging Fourier series for $\overline{B}_2(t)$ and appealing to a formula for the Mellin transform. We refer to the above references.

Proposition 5.3   *The product representation for the sine function*

$$\frac{\sin \pi z}{\pi z} = \prod_{n=1}^{\infty} \left(1 - \frac{z^2}{n^2} \right), \quad (5.58)$$

is a consequence of (5.49) and (5.53).

Proof.   Writing $-z$ for z in (5.49), we get

$$\Gamma(1-z) = e^{\gamma z} \prod_{n=1}^{\infty} e^{-\frac{z}{n}} \left(1 - \frac{z}{n} \right)^{-1}. \quad (5.59)$$

Multiplying (5.49) and (5.59), we deduce that

$$z \, \Gamma(z) \, \Gamma(1-z) = \prod_{n=1}^{\infty} \left(1 - \frac{z^2}{n^2} \right)^{-1}, \quad (5.60)$$

where we used the formula $\Gamma(z+1) = z\Gamma(z)$, which is also a consequence of (5.49) and (5.16). Plugging (5.53) in (5.60) gives (5.58).   □

Proposition 5.4   *The partial fraction expansion for the cotangent function*

$$\cot \pi z = \frac{1}{\pi z} - \frac{2z}{\pi} \sum_{n=1}^{\infty} \frac{1}{n^2 - z^2}$$

$$= \frac{1}{\pi z} + \frac{1}{\pi} \sum_{n=1}^{\infty} \left(\frac{1}{n+z} - \frac{1}{n-z} \right), \tag{5.61}$$

is a consequence of (5.58).

Proof.   This follows immediately by logarithmic differentiation. $\square$

Remark 5.6   *Comparing* (5.17) *and* (5.61), *we cover formula* (2.55) *again.*

Proposition 5.5   *The partial fraction expansion for the hyperbolic cotangent function and* (5.61) *are equivalent:*

$$\frac{1}{2} \coth \pi x = \frac{1}{e^{2\pi x} - 1} + \frac{1}{2} = \frac{1}{2\pi x} + \frac{x}{\pi} \sum_{n=1}^{\infty} \frac{1}{n^2 + x^2}, \quad \operatorname{Re} x \geq 0. \tag{5.62}$$

Proof.   This follows by putting $ix = z$ in (5.61) (i.e., we move from the right half-plane into the upper half-plane). $\square$

Lemma 5.4   *The partial fraction expansion for* $\coth x$ *and the (symmetric form) functional equation*

$$\pi^{-\frac{s}{2}} \Gamma\left(\frac{s}{2}\right) \zeta(s) = \pi^{-\frac{1-s}{2}} \Gamma\left(\frac{1-s}{2}\right) \zeta(1-s), \tag{5.63}$$

are equivalent.

Proof.   This can be found in [KTTY4], [Ko]. $\square$

Remark 5.7   *Historically,* (5.62) *was first used to deduce* (5.63) *(the fifth proof of [Tit], where an appeal to a formula (cf. Corollary A.4) for the Mellin transform is needed). Then Koshlyakov [Ko] deduced* (5.62) *from* (5.63).

Supplementarily, we state a result which allows us to skip above propositions and deduce (5.61) *directly from* (5.56) *or rather its equivalent under* (5.53):

$$l_{-s}(x) = i \frac{\Gamma(1+s)}{(2\pi)^{1+s}} \left\{ e^{\frac{\pi i s}{2}} \zeta(1+s, x) - e^{-\frac{\pi i s}{2}} \zeta(1+s, 1-x) \right\}. \tag{5.56$'$}$$

Proposition 5.6 *The functional equation* (5.56) *for the Hurwitz zeta-function implies the partial fraction expansion* (5.61) *for the* cot-*function.*

Proof. We remark that the functional equation (5.56) for $\zeta(s, x)$ may be expressed on the basis of (5.53) as (5.56)$'$.

First we assume that $\mathrm{Im}\, x > 0$. Then the sum for $l_0(x)$ converges for every $s \in \mathbb{C}$, and the left-hand side is

$$l_0(x) = \sum_{n=1}^{\infty} e^{2\pi inx} = \frac{e^{2\pi ix}}{1 - e^{2\pi ix}} = \frac{1}{2}\left(-1 + i \cot \pi x\right). \tag{5.64}$$

By analytic continuation, this holds true for every $x \in \mathbb{R} - \mathbb{Z}$.

We consider the limit as $s \to 0, s > 0$ on the right hand side of (5.56)$'$. First we note that

$$\left\{e^{\frac{\pi i s}{2}}\zeta(1 + s, x) - e^{-\frac{\pi i s}{2}}\zeta(1 + s, 1 - x)\right\} - \left\{\zeta(1 + s, x) - \zeta(1 + s, 1 - x)\right\}$$

$$= \left(e^{\frac{\pi i s}{2}} - 1\right)\zeta(1 + s, x) - \left(e^{-\frac{\pi i s}{2}} - 1\right)\zeta(1 + s, 1 - x)$$

$$= 2i \sin \frac{\pi s}{4}\left\{e^{\frac{\pi i s}{4}}\zeta(1 + s, x) + e^{-\frac{\pi i s}{4}}\zeta(1 + s, 1 - x)\right\}$$

$$= 2i \frac{\sin \frac{\pi s}{4}}{\frac{\pi}{4}s}\frac{\pi}{4}\left\{e^{\frac{\pi i s}{4}}s\zeta(1 + s, x) + e^{-\frac{\pi i s}{4}}s\zeta(1 + s, 1 - x)\right\},$$

which tends to πi as $s \to 0$ on account of $\lim_{s \to 0} s\,\zeta(1 + s, x) = 1$.

Secondly, since

$$\zeta(1 + s, x) - \zeta(1 + s, 1 - x) = \sum_{n=0}^{\infty}\left(\frac{1}{(n + x)^{1+s}} - \frac{1}{(n + 1 - x)^{1+s}}\right), \quad \sigma > 0$$

we get

$$\lim_{s \to 0, s > 0}\left(\zeta(1 + s, x) - \zeta(1 + s, 1 - x)\right)$$

$$= \sum_{n=0}^{\infty}\left(\frac{1}{n + x} - \frac{1}{n + 1 - x}\right) = \frac{1}{x} + \sum_{n=1}^{\infty}\frac{2x}{x^2 - n^2}.$$

Hence the limit of the right-hand side of (5.56)$'$ as $s \to 0$ through positive values is

$$i\frac{1}{x} + \sum_{n=1}^{\infty}\frac{2x}{x^2 - n^2}.$$

Combining this with (5.64), we conclude (5.61). $\qquad\qquad\square$

Proposition 5.7 *The functional equations in symmetric form* (5.63) *and in asymmetric form* (5.54) *are equivalent under* (5.48) *and* (5.53).

Proof is immediate.

Lemma 5.5 *The functional equation* (5.63) *for the Riemann zeta-function and* (5.56) *for the Hurwitz zeta-function are equivalent.*

Proof. This can be found in [KTTY4], [BKT] and is a manifestation of the most far-reaching modular relation principle. □

We are now in a position to state the main result of this chapter.

Theorem 5.5 *Under some known formulas, all formulas* (5.53), (5.54), (5.56), (5.58), (5.61), (5.62) *and* (5.63) *are equivalent in the sense of the following logical scheme (the portion including* (5.65) *is due to Theorem 5.6):*

$$(5.17) \implies (5.49)$$
$$(5.56) \Rightarrow (5.54) \Rightarrow (5.53)$$

$$(5.58) \Leftrightarrow (5.61) \Leftrightarrow (5.62) \Leftrightarrow (5.63) \Leftrightarrow (5.56)$$
$$\Updownarrow \qquad\qquad \Updownarrow$$
$$(5.65) \Leftarrow (5.56) \qquad\qquad (5.54)$$

Lemma 5.6 (Berndt) *The functional equation* (5.56) *for the Hurwitz zeta function implies Kummer's Fourier series for* $\log \Gamma(x)$, *which reads*

$$\log \frac{\Gamma(x)}{\sqrt{2\pi}} = -\frac{1}{2}\log(2\sin \pi x) + \frac{1}{2}(\gamma + \log 2\pi)(1 - 2x)$$
$$+ \frac{1}{\pi}\sum_{n=1}^{\infty}\frac{\log n}{n}\sin 2\pi nx, \tag{5.65}$$

which implies the reciprocal relation (5.49).

Proof is given by Berndt [Ber2], which depends on Lerch's formula (5.4) and the integral representation (5.1) for $\zeta(s, z)$ with $z = 0$.

Theorem 5.6 *Kummer's Fourier series for* $\log \Gamma(x)$ *is equivalent to the functional equation* (5.63) *for the Riemann zeta-function.*

Exercise 5.4 Deduce Euler's identity

$$\frac{B_{2m}}{(2m)!} = (-1)^{m-1}\frac{2\,\zeta(2m)}{(2\pi)^{2m}}, \quad m \geq 1 \tag{5.66}$$

from the partial fraction expansion (5.62) for the hyperbolic cotangent function.

Solution   Rewriting (5.66) in the form

$$\frac{2\pi x}{e^{2\pi x} - 1} + \pi x = 1 + 2x^2 \sum_{n=1}^{\infty} \frac{1}{n^2 + x^2}$$

and putting $2\pi x = z$, we obtain

$$\frac{z}{e^z - 1} = 1 - \frac{1}{2} z + 2z^2 \sum_{n=1}^{\infty} \frac{1}{z^2 + (2\pi n)^2} \qquad (5.67)$$

$$= 1 - \frac{1}{2} z + 2z^2 \varphi\left(z^2\right)$$

say, where

$$\varphi(w) = \sum_{n=1}^{\infty} \left(w + 4\pi^2 n^2\right)^{-1}. \qquad (5.68)$$

Since

$$\varphi^{(r)}(w) = (-1)^r\, r! \sum_{r=0}^{\infty} \left(w + 4\pi^2 n^2\right)^{-r-1},$$

we see that

$$\frac{\varphi^{(r)}(0)}{r!} = (-1)^r \sum_{n=1}^{\infty} \frac{1}{(2\pi n)^{2r+2}} = \frac{(-1)^r}{(2\pi)^{2r+2}} \zeta(2r + 2).$$

Hence

$$\frac{z}{e^z - 1} = 1 - \frac{1}{2} z + 2 \sum_{r=0}^{\infty} \frac{(-1)^r}{(2\pi)^{2r+2}} \zeta(2r + 2)\, z^{2r+2},$$

and so

$$\frac{z}{e^z - 1} = 1 - \frac{1}{2} z + \sum_{m=1}^{\infty} \frac{2\,(-1)^{m-1}}{(2\pi)^{2m}} \zeta(2m)\, z^{2m}.$$

Recalling the expansion

$$\frac{z}{e^z - 1} = 1 - \frac{1}{2} z + \sum_{m=1}^{\infty} \frac{B_{2m}}{(2m)!}\, z^{2m},$$

we conclude (5.66).

Remark 5.8 *As is proved above, (5.62) and the functional equation for the Riemann zeta-function are equivalent, whence we see that Euler's identity (5.66), and in particular the solution to the Basel problem $\zeta(2) = \frac{\pi^2}{6}$, is a consequence of the functional equation.*

Chapter 6

The theory of Bessel functions and the Epstein zeta-functions

Abstract

In this chapter we study an energy invariant – the Madelung constant associated to a crystal lattice through the lattice zeta-function, which is manifested as the Epstein zeta-function. We take into account the lattice structure (crystal symmetry) in our study through the functional equation of the Epstein zeta-function (zeta symmetry).

6.1 Introduction and the theory of Bessel functions

In this chapter we are going to study an energy invariant associated to a crystal lattice, called the Madelung constant about which numerous papers have appeared so far. Main references in book form are [Bor], [GZ] and [Ter1]. The main feature of our treatment in this regard is that we incorporate the lattice structure in its full extent, especially, the relationships between mutually dual lattice structures are revealed as those between the associated lattice zeta-functions, which are in turn manifested as the Epstein zeta-functions. That is, unlike previous work (save for Terras), we are going to express the distance and ion charges of the crystal structure in the form of a quadratic form and construct the Epstein zeta-functions associated to it, and then apply decomposition of the coefficient matrices to the Epstein zeta-function as is seen in Terras [Ter1]. For more details we refer to [KTTY2].

As the second main feature, we shall present a rather complete version of the theory of Epstein zeta-functions, which include generalizations of the theory of Berndt [Ber6], Chowla-Selberg and Terras as well as a unification of the theory of lattice zeta-values developed so far. They are manifested as the special values like $\zeta(1/2), \beta(1/2)$, where $\beta(s) = L(s, \chi_4)$ referred to in Abstract of Chapter 4, and we may efficiently incorporate our recent results on special values (see [KTY7]), using a perturbed Dirichlet series, or the Mellin-Barnes integrals (6.70) (cf. Paris-Kaminski [PK]).

Definition 6.1   The n-th **Bessel function** $J_n(z)$ is defined as the n-th Laurent coefficient of the function $\exp\left(\frac{z}{2}(w - \frac{1}{w})\right)$ in w, viz.

$$\exp\left(\frac{z}{2}\left(w - \frac{1}{w}\right)\right) = \sum_{n=-\infty}^{\infty} J_n(z) \, w^n. \tag{6.1}$$

Proposition 6.1   *We have the integral representation (called Bessel's integral)*

$$J_n(z) = \frac{1}{\pi} \int_0^{\pi} \cos(z \sin\theta - n\theta) \, \mathrm{d}\theta. \tag{6.2}$$

Proof.   By Theorem A.11, we have the integral representation

$$J_n(z) = \frac{1}{2\pi i} \int_{|w|=1} \frac{\exp\left(\frac{z}{2}(w - \frac{1}{w})\right)}{w^{n+1}} \, \mathrm{d}w.$$

By the parametric expression for the curve $|w| = 1$: $w = e^{i\theta}$, $0 \leq \theta \leq 2\pi$, we may rewrite the above as

$$J_n(z) = \frac{1}{2\pi} \int_0^{2\pi} e^{i(z \sin\theta - n\theta)} \, \mathrm{d}\theta \tag{6.3}$$

on noting that $w - \frac{1}{w} = 2i \sin\theta$. Dividing the interval $[0, 2\pi]$ into $[0, \pi]$ and $[\pi, 2\pi]$ and making the change of variable in the integral over $[\pi, 2\pi]$, we obtain

$$J_n(z) = \frac{1}{2\pi} \int_0^{\pi} e^{-i(z \sin\theta - n\theta)} \, \mathrm{d}\theta.$$

Adding this to the integral over $[0, \pi]$, we conclude (6.2). $\square$

Proposition 6.2   *The ν-th Bessel function $J_\nu(z)$ may be defined by*

$$J_\nu(z) = \sum_{n=0}^{\infty} \frac{(-1)^n}{n! \, \Gamma(\nu + n + 1)} \left(\frac{z}{2}\right)^{\nu + 2n}. \tag{6.4}$$

Proof. For $\nu \in \mathbb{Z}$, $J_\nu(z)$ is also defined as the ν-th coefficient of the product of two power series for $e^{\frac{z}{2}w}$ and $e^{-\frac{z}{2}\frac{1}{w}}$. Hence the ν-th term is given by

$$\sum_{m-n=\nu} \left(\frac{z}{2}\right)^\nu \frac{1}{m!} \frac{(-1)^n}{n!} = \left(\frac{z}{2}\right)^\nu \sum_{n=0}^\infty \frac{(-1)^n}{\Gamma(\nu+n+1)\,n!} \left(\frac{z}{2}\right)^2,$$

as claimed. For other values of ν, we understand (6.4) as the definition. $\square$

We note that from (6.2) it results

$$J_{-n}(z) = (-1)^n J_n(z) \tag{6.5}$$

Exercise 6.1 Viewing (6.1) as a Fourier series, deduce (6.2).

Solution With $w = e^{i\theta}$ ($\theta \in \mathbb{R}$), (6.1) reads

$$e^{iz\sin\theta} = \sum_{n=-\infty}^\infty J_n(z)\,e^{in\theta},$$

which is a Fourier series converging to the left-hand side member (in view of Theorem 7.2) and the Fourier coefficient $J_n(z)$ may be computed by (7.2):

$$J_n(z) = \frac{1}{2\pi} \int_{-\pi}^\pi e^{i(z\sin\theta - n\theta)}\,\mathrm{d}\theta,$$

which is (6.3), and we may argue as in the proof of Proposition 6.1.

Exercise 6.2 Use (6.5) to deduce for n even

$$\frac{1}{\pi} \int_0^\pi \cos(z\sin\theta)\cos(n\theta)\,\mathrm{d}\theta = J_n(z) \tag{6.6}$$

$$\frac{1}{\pi} \int_0^\pi \sin(z\sin\theta)\sin(n\theta)\,\mathrm{d}\theta = 0$$

while for n odd,

$$\frac{1}{\pi} \int_0^\pi \cos(z\sin\theta)\cos(n\theta)\,\mathrm{d}\theta = 0$$

$$\frac{1}{\pi} \int_0^\pi \sin(z\sin\theta)\sin(n\theta)\,\mathrm{d}\theta = J_n(z) \tag{6.7}$$

Solution $J_n(z) + J_{-n}(z)$ is $2J_n(z)$ for n even and 0 for n odd, and by (6.2), this is

$$\frac{1}{\pi} \int_0^\pi (\cos(z \sin \theta - n\theta) + \cos(z \sin \theta + n\theta))\, \mathrm{d}\theta,$$

whence the first and the second identities follow.

Considering $J_n(z) - J_{-n}(z)$, we deduce the third and the fourth identities.

In case $\nu \notin \mathbb{Z}$, the Bessel functions $J_\nu(x)$ and $J_{-\nu}(x)$ are two independent solutions to the Bessel differential equation

$$\frac{\mathrm{d}^2 y}{\mathrm{d}x^2} + \frac{1}{x}\frac{\mathrm{d}y}{\mathrm{d}x} + \left(1 - \frac{n^2}{x^2}\right) y = 0. \tag{6.8}$$

For $n \in \mathbb{Z}$, the fundamental solutions to (6.8) are given by $J_n(x)$ and $Y_n(x)$, the Weber function, relevant to analytic number theory. $J_\nu(z)$ and $Y_\nu(z)$ are often referred to as the **Bessel function of the first kind and of the second kind**, respectively.

Equally relevant to number theory are modified Bessel functions. The **modified Bessel function $I_\nu(z)$ of the first kind** is defined by

$$I_\nu(z) = \sum_{n=0}^\infty \frac{1}{n!\,\Gamma(\nu + n + 1)} \left(\frac{z}{2}\right)^{\nu + 2n}, \tag{6.9}$$

whence

$$I_n(z) = i^{-n} J_n(iz)$$

for $n \in \mathbb{Z}$.

The **modified Bessel function $K_\nu(z)$ of the second kind** is defined by

$$K_\nu(z) = \frac{\pi}{2} \frac{I_{-\nu}(z) - I_\nu(z)}{\sin \pi \nu} \tag{6.10}$$

(the limit is to be taken for $\nu \in \mathbb{Z}$), which has the integral representation

$$K_\nu(z) = \frac{1}{2} \int_0^\infty e^{-\frac{1}{2}z(t + \frac{1}{t})} t^{\nu - 1}\, \mathrm{d}t, \quad \mathrm{Re}\,\nu > -\frac{1}{2}, |\arg z| < \frac{\pi}{4}. \tag{6.11}$$

This appears in the proof of Theorem 6.1 in the context of Mellin inversion (sometimes referred to as the inverse Heaviside integral)

$$\frac{1}{2\pi i} \int_{(c)} \Gamma\left(s + \frac{\mu + \nu}{2}\right) \Gamma\left(s + \frac{\mu - \nu}{2}\right) x^{-s}\, \mathrm{d}s = 2\,x^{\frac{\mu}{2}} K_\nu(2\sqrt{x}), \tag{6.12}$$

for $c + \mathrm{Re}\,\frac{\mu+\nu}{2} \geq \mathrm{Re}\,\nu > 0$.

Bessel functions are, in a sense, generalizations of exponential functions, and they reduce to these functions for half-integral order, e.g.

$$J_{\frac{1}{2}}(z) = \sqrt{\frac{2}{\pi z}} \sin(z), \quad J_{-\frac{1}{2}}(z) = \sqrt{\frac{2}{\pi z}} \cos(z) \qquad (6.13)$$

and

$$K_{\frac{1}{2}}(z) = K_{-\frac{1}{2}}(z) = \sqrt{\frac{\pi}{2z}}\, e^{-z}. \qquad (6.14)$$

6.2 The theory of Epstein zeta-functions

We now introduce the notation (from Terras [Ter1]) concerning the Epstein zeta-functions, which will be used throughout in what follows.

Notation. Let g, $h \in \mathbb{R}^n$ be n-dimensional real vectors which (in the first place) give rise to the perturbation and the (additive) characters, respectively.

Let $Y = (y_{ij})$ be a positive definite $n \times n$ real symmetric matrix. Define the Epstein zeta-function associated to the quadratic form

$$Y[a] = a \cdot Ya = {}^t aYa = \sum_{i,j=1}^{n} y_{ij} a_i a_j, \qquad (6.15)$$

where $a = (a_1, \ldots, a_n) \in \mathbb{R}^n$ and "$\cdot$" means the scalar product, by

$$Z(Y, 0, 0, s) = \sum_{\substack{a \in \mathbb{Z}^n \\ a \neq 0}} \frac{1}{Y[a]^s}, \quad \sigma > \frac{n}{2}, \qquad (6.16)$$

where $\sigma = \mathrm{Re}\,s$.

For g, $h \in \mathbb{R}^n$ define the general **Epstein zeta-function** (of Hurwitz-Lerch type) by

$$Z(Y, g, h, s) = \sum_{\substack{a \in \mathbb{Z}^n \\ a+g \neq 0}} \frac{e^{2\pi i h \cdot a}}{Y[a+g]^s}, \quad \sigma > \frac{n}{2}, \qquad (6.17)$$

and incorporate the **completion**

$$\Lambda(Y, g, h, s) = \pi^{-s}\Gamma(s)\, Z(Y, g, h, s), \qquad (6.18)$$

which satisfies the functional equation of the form (5.63) with an additional factor and replacement of parameters (proof given in §6.4):

$$\Lambda(Y, g, h, s) = \frac{1}{\sqrt{|Y|}} e^{-2\pi i g \cdot h} \Lambda\left(Y^{-1}, h, -g, \frac{n}{2} - s\right). \qquad (6.19)$$

In what follows we always denote the special vector ${}^t\left(\frac{1}{2}, \frac{1}{2}, \frac{1}{2}\right)$ by c_0:

$$c_0 = \begin{pmatrix} \frac{1}{2} \\ \frac{1}{2} \\ \frac{1}{2} \end{pmatrix}.$$

We shall now give some illustrative examples.

Example 6.1   The relationship between the Madelung constants of the *NaCl* and *CsCl* structure.

In [Hautot, p.1724], it is stated that the cations of *CsCl* are at $a \in \left(\frac{2}{\sqrt{3}}\left(\mathbb{Z} + \frac{1}{2}\right)\right)^3$ and anions are at $a \in \left(\frac{2}{\sqrt{3}}\mathbb{Z}\right)^3$. The Madelung constant M_{CsCl} is defined, in the first place, by

$$M_{CsCl} = \frac{\sqrt{3}}{2} \sum_{a \in \mathbb{Z}^3} |a + c_0|^{-1} - \frac{\sqrt{3}}{2} \sum_{\substack{a \in \mathbb{Z}^3 \\ a \neq 0}} |a|^{-1}, \qquad (6.20)$$

which, in our notation above, is

$$\frac{\sqrt{3}}{2} Z\left(I, c_0, 0, \frac{1}{2}\right) - \frac{\sqrt{3}}{2} Z\left(I, 0, 0, \frac{1}{2}\right) \qquad (6.21)$$

and is in turn equal to

$$-\sqrt{3} Z\left(B, 0, c_0, \frac{1}{2}\right), \qquad (6.22)$$

where

$$I = \begin{pmatrix} 1 & 0 & 0 \\ 0 & 1 & 0 \\ 0 & 0 & 1 \end{pmatrix} \quad \text{(identity matrix)},$$

and

$$B = \begin{pmatrix} 3 & -1 & -1 \\ -1 & 3 & -1 \\ -1 & -1 & 3 \end{pmatrix}. \qquad (6.23)$$

Hautot [Hautot], without giving any reasons, transforms (6.20) into the form

$$\frac{2}{\sqrt{3}} M_{CsCl} = 2M_{NaCl}$$

$$+ 6 \sum \sum \sum \left[\left\{ (2l)^2 + (2m+1)^2 + (2n+1)^2 \right\}^{-1/2} \right. \tag{6.24}$$

$$\left. - \left\{ (2l)^2 + (2m+1)^2 + (2n)^2 \right\}^{-1/2} \right],$$

and then proceeds to transfer the triple sum using the Schlömilch series technique (cf. [KTZ] also). Thus (6.24) suggests that there may be a relationship between M_{NaCl} and M_{CsCl} structure. This suggestion is strengthened by the comparison of numerical values

$$M_{NaCl} = 1.74756459463\ldots,$$
$$M_{CsCl} = 1.76267477307\ldots. \tag{6.25}$$

The real situation is the following duality relations (6.26) and (6.27), which can be found only through the study of lattice structures.

Between the Madelung constants $M_{NaCl} = -Z(I, \mathbf{0}, c_0, \frac{1}{2})$ and $M_{CsCl} = -\sqrt{3}\, Z(B, \mathbf{0}, c_0, \frac{1}{2})$, the duality relations hold (under the notation (6.23)):

$$M_{NaCl} = -Z\left(I, \mathbf{0}, c_0, \frac{1}{2}\right) = -\frac{2}{\pi} \left\{ Z(B, \mathbf{0}, \mathbf{0}, 1) - Z(B, \mathbf{0}, c_0, 1) \right\} \tag{6.26}$$

and

$$M_{CsCl} = -\sqrt{3}\, Z\left(B, \mathbf{0}, c_0, \frac{1}{2}\right) = -\frac{\sqrt{3}}{2\pi} \left\{ Z(I, \mathbf{0}, \mathbf{0}, 1) - Z(I, \mathbf{0}, c_0, 1) \right\} \tag{6.27}$$

(cf. Formula (1.8) on p.721 of [KTTY1]; proof given in Example 6.2 below).

In the case of M_{ZnS}, Hautot states another relation corresponding to (6.24), again without giving any reason why the Madelung constants M_{ZnS} and M_{CsCl} should be related:

$$\frac{4}{\sqrt{3}} M_{ZnS} = \frac{2}{\sqrt{3}} M_{CsCl} - 6 \sum \sum \sum \frac{1}{\sqrt{(2l)^2 + (2m+1)^2 + (2n+1)^2}}, \tag{6.28}$$

where we note that the Madelung constant M_{ZnS} is to be defined by

$$M_{ZnS} = \frac{\sqrt{3}}{2} Z\left(A, \frac{1}{2}c_0, \mathbf{0}, \frac{1}{2}\right) - \frac{\sqrt{3}}{2} Z\left(A, \mathbf{0}, \mathbf{0}, \frac{1}{2}\right), \tag{6.29}$$

where

$$A = \begin{pmatrix} 2 & 1 & 1 \\ 1 & 2 & 1 \\ 1 & 1 & 2 \end{pmatrix}. \tag{6.30}$$

Comparing (6.27) and (6.29) and the numerical values (6.25) and (6.31) below does not give much to expect a relation between them;

$$M_{ZnS} = 1.63805505338\ldots. \tag{6.31}$$

Surprisingly enough, there holds a remarkable relation

$$M_{ZnS} = \frac{\sqrt{3}}{4} M_{NaCl} + \frac{1}{2} M_{CsCl}. \tag{6.32}$$

For a proof see [KTTY2].

In Example 6.2 we are going to reveal those identities given in Example 6.1 as special cases of zeta-function relations.

We introduce the general principle.

Principle. Suppose L is a lattice with basis e_1, e_2, e_3:

$$L = \mathbb{Z}e_1 \oplus \mathbb{Z}e_2 \oplus \mathbb{Z}e_3. \tag{6.33}$$

With $M = (e_1, e_2, e_3)$, the associated Gram matrix Y is defined by ${}^t M M$:

$$Y = {}^t M M = \begin{pmatrix} e_1 \cdot e_1 & e_1 \cdot e_2 & e_1 \cdot e_3 \\ e_2 \cdot e_1 & e_2 \cdot e_2 & e_2 \cdot e_3 \\ e_3 \cdot e_1 & e_3 \cdot e_2 & e_3 \cdot e_3 \end{pmatrix}. \tag{6.34}$$

Let $f_1 = e_2 + e_3$, $f_2 = e_3 + e_1$, $f_3 = e_1 + e_2$, and

$$J = \begin{pmatrix} 0 & 1 & 1 \\ 1 & 0 & 1 \\ 1 & 1 & 0 \end{pmatrix}. \tag{6.35}$$

Then the matrix ${}^t J Y J = J Y J$ is the Gram matrix associated to the sublattice $L_1 = \mathbb{Z}f_1 \oplus \mathbb{Z}f_2 \oplus \mathbb{Z}f_3$ of L, and we have

$$L = L_1 \cup \left(L_1 + \frac{1}{2} f_1 + \frac{1}{2} f_2 + \frac{1}{2} f_3 \right). \tag{6.36}$$

We appeal to the fact to be proved in §6.3 (Proposition 6.3) that the zeta-function of a lattice coincides with the Epstein zeta-function of the corresponding Gram matrix under suitable identification. It follows that

$$\Lambda(Y, \mathbf{0}, \mathbf{c_0}, s) = \Lambda({}^t J\, Y J, \mathbf{0}, \mathbf{0}, s) - \Lambda({}^t J\, Y J, \mathbf{c_0}, \mathbf{0}, s)\,, \tag{6.37}$$

and

$$\Lambda(Y, \mathbf{0}, \mathbf{0}, s) = \Lambda({}^t J\, Y J, \mathbf{0}, \mathbf{0}, s) + \Lambda({}^t J\, Y J, \mathbf{c_0}, \mathbf{0}, s)\,. \tag{6.37}'$$

Now note that the inverse matrix $({}^t J\, Y J)^{-1}$ is the Gram matrix associated to the dual lattice $L_1'(\cong \mathrm{Hom}(L_1, \mathbb{Z}))$ or recall the functional equation (6.19) to transform the right-hand side of (6.37) further into

$$\frac{1}{\sqrt{|{}^t J\, Y J|}}\left\{\Lambda\left(({}^t J\, Y J)^{-1}, \mathbf{0}, \mathbf{0}, \frac{3}{2} - s\right) - \Lambda\left(({}^t J\, Y J)^{-1}, \mathbf{0}, -\mathbf{c_0}, \frac{3}{2} - s\right)\right\},$$

so that

$$\Lambda(Y, \mathbf{0}, \mathbf{c_0}, s)$$
$$= \frac{1}{\sqrt{|{}^t J\, Y J|}}\left\{\Lambda\left(({}^t J\, Y J)^{-1}, \mathbf{0}, \mathbf{0}, \frac{3}{2} - s\right) - \Lambda\left(({}^t J\, Y J)^{-1}, \mathbf{0}, -\mathbf{c_0}, \frac{3}{2} - s\right)\right\}. \tag{6.38}$$

Now we apply the above principle to some lattice sums.

Example 6.2   First choose ${}^t J\, Y J = A = \begin{pmatrix} 2 & 1 & 1 \\ 1 & 2 & 1 \\ 1 & 1 & 2 \end{pmatrix}$ ((6.30)). Then $Y = I$

and (6.37) reads

$$Z(I, \mathbf{0}, \mathbf{c_0}, s) = Z(A, \mathbf{0}, \mathbf{0}, s) - Z(A, \mathbf{c_0}, \mathbf{0}, s). \tag{6.39}$$

This explains the reason why the proper definition (given in [KTTY2]) of the Madelung constant M_{NaCl} as the value at $s = \frac{1}{2}$ of

$$Z(A, \mathbf{c_0}, \mathbf{0}, s) - Z(A, \mathbf{0}, \mathbf{0}, s)$$

coincides with the value at $s = \frac{1}{2}$ of $-Z(I, \mathbf{0}, \mathbf{c_0}, s)$ i.e.

$$M_{NaCl} = Z\left(A, \mathbf{c_0}, \mathbf{0}, \frac{1}{2}\right) - Z\left(A, \mathbf{0}, \mathbf{0}, \frac{1}{2}\right) = -Z\left(I, \mathbf{0}, \mathbf{c_0}, \frac{1}{2}\right). \tag{6.40}$$

Next, we choose $Y = \frac{1}{4}B$ ((6.23)). Then ${}^t J\, Y J = I$, and

$$Z\left(\frac{1}{4}B, \mathbf{0}, \mathbf{c_0}, s\right) = Z(I, \mathbf{0}, \mathbf{0}, s) - Z(I, \mathbf{c_0}, \mathbf{0}, s). \tag{6.41}$$

We make an important remark, which will be in effect in treating the Abel mean in [KTTY2], that for $c > 0$

$$Z(cY, \boldsymbol{g}, \boldsymbol{h}, s) = c^{-s} Z(Y, \boldsymbol{g}, \boldsymbol{h}, s), \tag{6.42}$$

i.e. we may incorporate the parameter c in Y by just multiplying by the factor c^{-s}.

Using (6.42) and (6.41), we have for $s = \frac{1}{2}$

$$\frac{\sqrt{3}}{2} \left\{ Z\left(I, \boldsymbol{c}_0, \boldsymbol{0}, \frac{1}{2}\right) - Z\left(I, \boldsymbol{0}, \boldsymbol{0}, \frac{1}{2}\right) \right\} = -\sqrt{3}\, Z\left(B, \boldsymbol{0}, \boldsymbol{c}_0, \frac{1}{2}\right), \tag{6.43}$$

which asserts that (6.21) and (6.22) are equal.

We turn to the proof of duality relations (6.26) and (6.27).

As we deduced (6.39), we choose ${}^t\!J Y J = A$, and so $Y = I$. Then (6.38) gives

$$-\Lambda(I, \boldsymbol{0}, \boldsymbol{c}_0, s) = \frac{1}{\sqrt{|A|}} \left\{ \Lambda\left(A^{-1}, \boldsymbol{0}, -\boldsymbol{c}_0, \frac{3}{2} - s\right) - \Lambda\left(A^{-1}, \boldsymbol{0}, \boldsymbol{0}, \frac{3}{2} - s\right) \right\}.$$

Since $A^{-1} = \frac{1}{4}B$, we apply (6.42) to obtain

$$-\Lambda(I, \boldsymbol{0}, \boldsymbol{c}_0, s) = \frac{1}{2}\, 4^{\frac{3}{2} - s} \left\{ \Lambda\left(B, \boldsymbol{0}, -\boldsymbol{c}_0, \frac{3}{2} - s\right) - \Lambda\left(B, \boldsymbol{0}, \boldsymbol{0}, \frac{3}{2} - s\right) \right\},$$

or

$$- \pi^{-s}\Gamma(s)\, Z(I, \boldsymbol{0}, \boldsymbol{c}_0, s) \tag{6.44}$$
$$= 2^{2-2s}\pi^{-\left(\frac{3}{2}-s\right)} \Gamma\left(\frac{3}{2} - s\right) \left\{ Z\left(B, \boldsymbol{0}, -\boldsymbol{c}_0, \frac{3}{2} - s\right) - Z\left(B, \boldsymbol{0}, \boldsymbol{0}, \frac{3}{2} - s\right) \right\},$$

which in turn gives (6.26) for $s = \frac{1}{2}$.

Also, for the choice $Y = \frac{1}{4}B$, ${}^t\!J Y J = I$, (6.38) reads, as in (6.43),

$$\Lambda(B, \boldsymbol{0}, \boldsymbol{c}_0, s) = 4^{-s} \left\{ \Lambda\left(I, \boldsymbol{0}, \boldsymbol{0}, \frac{3}{2} - s\right) - \Lambda\left(I, \boldsymbol{0}, -\boldsymbol{c}_0, \frac{3}{2} - s\right) \right\},$$

or

$$\pi^{-s}\Gamma(s)\, Z(B, \boldsymbol{0}, \boldsymbol{c}_0, s) \tag{6.45}$$
$$= 4^{-s}\pi^{-\left(\frac{3}{2}-s\right)} \Gamma\left(\frac{3}{2} - s\right) \left\{ Z\left(I, \boldsymbol{0}, \boldsymbol{0}, \frac{3}{2} - s\right) - Z\left(I, \boldsymbol{0}, -\boldsymbol{c}_0, \frac{3}{2} - s\right) \right\},$$

which gives (6.27) for $s = \frac{1}{2}$.

6.3 Lattice zeta-functions

In this section we shall clarify the relationship between the zeta-functions mentioned in the title and apply to the study of Madelung constants.

Let L be a lattice, i.e. a free Abelian group of finite rank (n, say) with biadditive form $(\ ,\ )_L$. We form the zeta-function $Z_L(s) = Z(L, 0, 0, s)$ corresponding to (6.16) by

$$Z_L(s) = Z(L, 0, 0, s) = \sum_{\substack{x \in L \\ x \neq 0}} \frac{1}{(x, x)_L^s}, \tag{6.46}$$

absolutely convergent for $\sigma > \frac{n}{2}$.

If, in particular, $L \subset \mathbb{R}^m$ and $(\boldsymbol{a}, \boldsymbol{b})_L$ means the scalar product $\boldsymbol{a} \cdot \boldsymbol{b} = {}^t\boldsymbol{a}\boldsymbol{b} = \sum_{i=1}^m a_i b_i$, then

$$Z(L, 0, 0, s) = \sum_{\substack{x \in L \\ x \neq 0}} \frac{1}{(x_1^2 + \cdots + x_m^2)^s}. \tag{6.47}$$

As usual, let L' denote the dual lattice of L: $L' = \mathrm{Hom}(L, \mathbb{Z})$. Then for lattice elements p, q with real coefficients, $p \in L \otimes \mathbb{R}$, $q \in L' \otimes \mathbb{R}$, we introduce the general lattice zeta-function $Z(L, p, q, s)$ corresponding to (6.17) by

$$Z(L, p, q, s) = \sum_{\substack{x \in L \\ x+p \neq 0}} \frac{e^{2\pi i q(x)}}{(x + p, x + p)_{L \otimes \mathbb{R}}^s}, \tag{6.48}$$

absolutely convergent for $\sigma > \frac{n}{2}$. Here we understand the meaning of $q(x)$ through

$$L' \otimes \mathbb{R} \cong \mathrm{Hom}(L, \mathbb{R}) \cong \mathrm{Hom}_{\mathbb{R}}(L \otimes \mathbb{R}, \mathbb{R})$$

and the completion corresponding to (6.18):

$$\Lambda(L, p, q, s) = \pi^{-s} \Gamma(s) \, Z(L, p, q, s). \tag{6.49}$$

We recall the Principle in §1 in the following form.

Associated to a lattice L with basis $e_1, \ldots, e_n$, $L = \mathbb{Z}e_1 \oplus \cdots \oplus \mathbb{Z}e_n$, is its Gram matrix

$$Y = {}^tMM = \begin{pmatrix} (e_1 \cdot e_1)_L & \cdots & (e_1 \cdot e_n)_L \\ \vdots & \ddots & \vdots \\ (e_n \cdot e_1)_L & \cdots & (e_n \cdot e_n)_L \end{pmatrix}, \tag{6.50}$$

where $M = (e_1, \ldots, e_n)$.

Let ϕ be the canonical isomorphism

$$\phi : \mathbb{Z}^n \longrightarrow L, \quad x = \phi(a) = a_1 e_1 + \cdots + a_n e_n, \qquad (6.51)$$

for $a = (a_1, \ldots, a_n)$, or $\phi(a) = Ma$. Through ϕ, we may interpret the bilinear form $(x, x)_L$ as $(\phi(a), \phi(a))_L$, which we may think of as $Y[a]$. Thus,

$$Z_L(s) = Z(L, 0, 0, s) = Z(Y, \mathbf{0}, \mathbf{0}, s). \qquad (6.52)$$

We may extend ϕ to the isomorphism

$$\phi : \mathbb{R}^n \longrightarrow L \otimes \mathbb{R}, \quad x = \phi(a) = a_1 e_1 + \cdots + a_n e_n,$$

for $a = (a_1, \ldots, a_n) \in \mathbb{R}^n$. Then we have

$$(\phi(a), \phi(a))_{L \otimes \mathbb{R}} = Y[a].$$

If further we put $p = \phi(g)$ and $q(x) = q \circ \phi(a) = h \cdot a \ (a \in \mathbb{R}^n)$, then

$$Z(L, p, q, s) = \sum_{\substack{a \in \mathbb{Z}^n \\ a + g \neq 0}} \frac{e^{2\pi i q \circ \phi(a)}}{(\phi(a + g), \phi(a + g))_{L \otimes \mathbb{R}}^s} \qquad (6.53)$$

$$= \sum_{\substack{a \in \mathbb{Z}^n \\ a + g \neq 0}} \frac{e^{2\pi i h \cdot a}}{Y[a + g]^s} = Z(Y, g, h, s).$$

Thus, we have

Proposition 6.3 *Under above notation, we have*

$$\Lambda(L, p, q, s) = \Lambda(Y, g, h, s).$$

Hence, whenever we speak about a lattice zeta-function, we may do well with the corresponding Epstein zeta-function with the Gram matrix.

Example 6.3 (i) The simple cubic (s.c.) structure, $\mathbb{Z}^3 = \mathbb{Z} \begin{pmatrix} 1 \\ 0 \\ 0 \end{pmatrix} \oplus$

$$\mathbb{Z}\begin{pmatrix} 0 \\ 1 \\ 0 \end{pmatrix} \oplus \mathbb{Z}\begin{pmatrix} 0 \\ 0 \\ 1 \end{pmatrix} \text{ with Gram matrix } I = \begin{pmatrix} 1\ 0\ 0 \\ 0\ 1\ 0 \\ 0\ 0\ 1 \end{pmatrix}. \text{ The zeta-function is}$$

$$Z\left(\mathbb{Z}^3, 0, 0, s\right) = Z(I, 0, 0, s) = \sum_{\substack{a \in \mathbb{Z}^3 \\ a \neq 0}} \frac{1}{|a|^{2s}}. \tag{6.54}$$

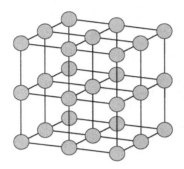

Fig. 6.1   the simple cubic (s.c.) structure

(ii) The face-centered cubic structure (f.c.c.),

$$L_f = \left\{ a \in \mathbb{Z}^3 \,\middle|\, (-1)^{a_1+a_2+a_3} = 1 \right\} = \mathbb{Z}\begin{pmatrix} 0 \\ 1 \\ 1 \end{pmatrix} \oplus \mathbb{Z}\begin{pmatrix} 1 \\ 0 \\ 1 \end{pmatrix} \oplus \mathbb{Z}\begin{pmatrix} 1 \\ 1 \\ 0 \end{pmatrix}$$

with Gram matrix $A = \begin{pmatrix} 2\ 1\ 1 \\ 1\ 2\ 1 \\ 1\ 1\ 2 \end{pmatrix}$ ((6.30)).

The zeta-function is

$$Z_{L_f}(s) = Z(A, 0, 0, s)$$
$$= \sum_{\substack{a \in \mathbb{Z}^3 \\ a \neq 0}} \frac{1}{(2a_1^2 + 2a_2^2 + 2a_3^2 + 2a_1a_2 + 2a_2a_3 + 2a_3a_1)^s}.$$

With $c_0 = {}^t(\frac{1}{2}, \frac{1}{2}, \frac{1}{2})$, $Z(I, 0, c_0, s) = \sum_{\substack{a \in \mathbb{Z}^3 \\ a \neq 0}} \frac{(-1)^{a_1+a_2+a_3}}{I[a]^s}$ can be written as

$$Z(I, 0, c_0, s) = 2Z(A, 0, 0, s) - Z(I, 0, 0, s).$$

Solving in $Z(A, \mathbf{0}, \mathbf{0}, s)$, we have

$$Z(A, \mathbf{0}, \mathbf{0}, s) = \frac{1}{2}\left(Z(I, \mathbf{0}, \mathbf{0}, s) + Z(I, \mathbf{0}, c_0, s)\right). \tag{6.55}$$

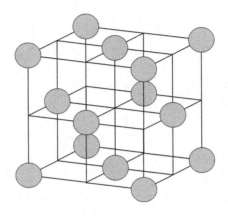

Fig. 6.2   the face-centered cubic structure (f.c.c.)

(iii) The body-centered cubic structure (b.c.c.),

$$L_b = \left\{ a \in \mathbb{Z}^3 \,|\, a_2 + a_3, a_3 + a_1, a_1 + a_2 \in 2\mathbb{Z} \right\}$$
$$= \left\{ a \in \mathbb{Z}^3 \,|\, (-1)^{a_2+a_3} + (-1)^{a_3+a_1} + (-1)^{a_1+a_2} = 3 \right\}$$

with Gram matrix $B = \begin{pmatrix} 3 & -1 & -1 \\ -1 & 3 & -1 \\ -1 & -1 & 3 \end{pmatrix}$ ((6.23)). The zeta-function is

$$Z_{L_b}(s) = Z(B, \mathbf{0}, \mathbf{0}, s)$$
$$= \sum_{\substack{a \in \mathbb{Z}^3 \\ a \neq 0}} \frac{1}{(3a_1^2 + 3a_2^2 + 3a_3^2 - 2a_1a_2 - 2a_2a_3 - 2a_3a_1)^s}.$$

Since

$$3\,Z\!\left(I,\mathbf{0},\begin{pmatrix}\frac{1}{2}\\\frac{1}{2}\\0\end{pmatrix},s\right) = Z\!\left(I,\mathbf{0},\begin{pmatrix}0\\\frac{1}{2}\\\frac{1}{2}\end{pmatrix},s\right) + Z\!\left(I,\mathbf{0},\begin{pmatrix}\frac{1}{2}\\0\\\frac{1}{2}\end{pmatrix},s\right) + Z\!\left(I,\mathbf{0},\begin{pmatrix}\frac{1}{2}\\\frac{1}{2}\\0\end{pmatrix},s\right)$$

$$= \sum_{\substack{a\in\mathbb{Z}^3\\a\neq0}} \frac{(-1)^{a_2+a_3} + (-1)^{a_3+a_1} + (-1)^{a_1+a_2}}{I[a]^s},$$

we get, on resorting to the definition of L_b,

$$3\,Z\!\left(I,\mathbf{0},\begin{pmatrix}\frac{1}{2}\\\frac{1}{2}\\0\end{pmatrix},s\right) = \sum_{\substack{a\in L_b\\a\neq0}} \frac{3}{I[a]^s} + \sum_{a\in\mathbb{Z}^3-L_b} \frac{-1}{I[a]^s}$$

$$= 4\,Z(B,\mathbf{0},\mathbf{0},s) - Z(I,\mathbf{0},\mathbf{0},s),$$

whence

$$Z(B,\mathbf{0},\mathbf{0},s) = \frac{1}{4}\left\{ Z(I,\mathbf{0},\mathbf{0},s) + 3\,Z\!\left(I,\mathbf{0},\begin{pmatrix}\frac{1}{2}\\\frac{1}{2}\\0\end{pmatrix},s\right)\right\}. \tag{6.56}$$

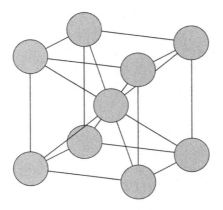

Fig. 6.3   the body-centered cubic structure (b.c.c.)

Using 'square root' of A and B, i.e. J from (6.35) and

$$K = \begin{pmatrix} -1 & 1 & 1 \\ 1 & -1 & 1 \\ 1 & 1 & -1 \end{pmatrix} \tag{6.57}$$

($J^2 = A$, $K^2 = B$, $J^{-1} = \frac{1}{2}K$), we obtain generalizations of formulas (6.55) and (6.56):

$$Z(JYJ, \boldsymbol{g}, \boldsymbol{h}, s) \tag{6.58}$$
$$= \frac{1}{2}\left\{ Z(Y, J\boldsymbol{g}, J^{-1}\boldsymbol{h}, s) + Z\left(Y, J\boldsymbol{g}, J^{-1}\boldsymbol{h} + \begin{pmatrix} \frac{1}{2} \\ \frac{1}{2} \\ \frac{1}{2} \end{pmatrix}, s\right) \right\}$$

$$Z(KYK, \boldsymbol{g}, \boldsymbol{h}, s) \tag{6.59}$$
$$= \frac{1}{4}\left\{ Z(Y, K\boldsymbol{g}, K^{-1}\boldsymbol{h}, s) + Z\left(Y, K\boldsymbol{g}, K^{-1}\boldsymbol{h} + \begin{pmatrix} 0 \\ \frac{1}{2} \\ \frac{1}{2} \end{pmatrix}, s\right) \right.$$
$$\left. + Z\left(Y, K\boldsymbol{g}, K^{-1}\boldsymbol{h} + \begin{pmatrix} \frac{1}{2} \\ 0 \\ \frac{1}{2} \end{pmatrix}, s\right) + Z\left(Y, K\boldsymbol{g}, K^{-1}\boldsymbol{h} + \begin{pmatrix} \frac{1}{2} \\ \frac{1}{2} \\ 0 \end{pmatrix}, s\right) \right\}.$$

Example 6.4 For the notation and more details, cf. [KTTY2].
(i) The $NaCl$ (Sodium Chloride) structure. Here the data is

$$n_+ = n_- = 1,$$
$$S_{++} = S_{--} = \{\boldsymbol{a} \in \mathbb{Z}^3 | a_1 + a_2 + a_3 \in 2\mathbb{Z}\} \text{ (f.c.c.)},$$
$$S_{+-} = S_{+-} = \{\boldsymbol{a} \in \mathbb{Z}^3 | a_1 + a_2 + a_3 \in 2\mathbb{Z} + 1\} \text{ (f.c.c.)},$$

so that by (6.40)

$$Z_{NaCl}(s) = Z(A, \boldsymbol{c}_0, \boldsymbol{0}, s) - Z(A, \boldsymbol{0}, \boldsymbol{0}, s) \tag{6.60}$$
$$= -Z(I, \boldsymbol{0}, \boldsymbol{c}_0, s).$$

Formula (6.60) justifies the first equality in (6.26).

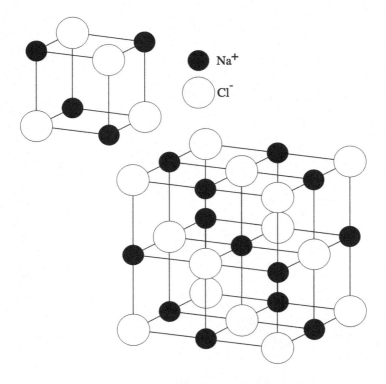

Fig. 6.4   the $NaCl$ (Sodium Chloride) structure (s.c.)

Thus, by (6.40), the Madelung constant M_{NaCl} is given by

$$M_{NaCl} = -Z\left(I, \mathbf{0}, c_0, \frac{1}{2}\right) \tag{6.61}$$

$$= -\sum_{\substack{\mathbf{a} \in \mathbb{Z}^3 \\ \mathbf{a} \neq \mathbf{0}}} \frac{(-1)^{a_1+a_2+a_3}}{|\mathbf{a}|} = 1.7475645849\ldots$$

as stated in (6.25) and (6.26).

(ii) The $CsCl$ (Caesium Chloride) structure. Here the data is

$$n_+ = n_- = 1,$$

$$S_{++} = S_{--} = \left(\frac{2}{\sqrt{3}}\mathbb{Z}\right)^3 \text{ (s.c.)},$$

$$S_{+-} = S_{+-} = \left(\frac{2}{\sqrt{3}}\left(\mathbb{Z}+\frac{1}{2}\right)\right)^3 \text{ (s.c.)},$$

and the zeta-function is, as discussed in Example 6.2, (6.41)–(6.43),

$$Z_{CsCl}(s) = Z\left(\frac{4}{3}I, \mathbf{c}_0, \mathbf{0}, s\right) - Z\left(\frac{4}{3}I, \mathbf{0}, \mathbf{0}, s\right) \qquad (6.62)$$

$$= \left(\frac{3}{4}\right)^s Z(I, \mathbf{c}_0, \mathbf{0}, s) - \left(\frac{3}{4}\right)^s Z(I, \mathbf{0}, \mathbf{0}, s)$$

$$= -3^s Z(B, \mathbf{0}, \mathbf{c}_0, s).$$

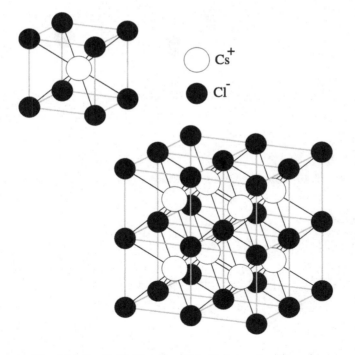

Fig. 6.5   the $CsCl$ (Caesium Chloride) structure (b.c.c.)

Hence

$$M_{CsCl} = \frac{\sqrt{3}}{2} Z\left(I, c_0, 0, \frac{1}{2}\right) - \frac{\sqrt{3}}{2} Z\left(I, 0, 0, \frac{1}{2}\right) \qquad (6.63)$$
$$= -\sqrt{3}\, Z\left(B, 0, c_0, \frac{1}{2}\right)$$

as in (6.43), whence (6.27) ensues.

(iii) The ZnS (Zincblende) structure. The data is

$$n_+ = n_- = 1,$$
$$S_{++} = S_{--} = \left\{ \frac{2}{\sqrt{3}} a \;\middle|\; a \in \mathbb{Z}^3, a_1 + a_2 + a_3 \in 2\mathbb{Z} \right\} \text{(f.c.c.)},$$
$$S_{+-} = S_{+-} = \left\{ \frac{2}{\sqrt{3}} a \;\middle|\; a \in \left(\mathbb{Z} + \frac{1}{2}\right)^3, a_1 + a_2 + a_3 \in 2\mathbb{Z} + \frac{1}{2} \right\} \text{(f.c.c.)},$$

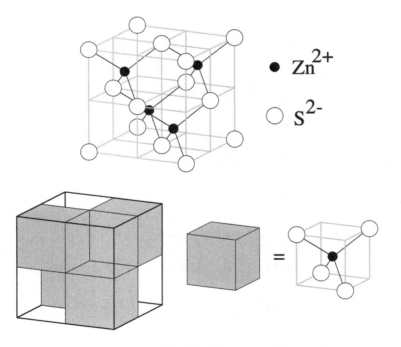

Fig. 6.6 the ZnS (Zincblende) structure (diamond)

and the zeta-function is

$$Z_{ZnS}(s) = Z\left(\frac{4}{3}A, \frac{1}{2}c_0, 0, s\right) - Z\left(\frac{4}{3}A, 0, 0, s\right) \tag{6.64}$$
$$= \left(\frac{3}{4}\right)^s Z\left(A, \frac{1}{2}c_0, 0, s\right) - \left(\frac{3}{4}\right)^s Z(A, 0, 0, s),$$

i.e. (6.29). Further in a similarly way as we prove (6.67) in [KTTY2], we may prove

$$Z_{ZnS}(s) = \frac{1}{2}\left(\frac{3}{4}\right)^s Z_{NaCl}(s) + \frac{1}{2}Z_{CsCl}(s), \tag{6.65}$$

whence as in (6.32) and (6.31)

$$M_{ZnS} = \frac{\sqrt{3}}{4}M_{NaCl} + \frac{1}{2}M_{CsCl} \tag{6.66}$$
$$= 1.63805805338\ldots.$$

(iv) The CaF_2 (Fluorite) structure. The data:

$n_+ = 1, n_- = 2,$

$$S_{++} = \left\{\frac{2}{\sqrt{3}}a \,\middle|\, a \in \mathbb{Z}^3, a_1 + a_2 + a_3 \in 2\mathbb{Z}\right\} \text{ (f.c.c.),}$$

$$S_{+-} = \left(\frac{2}{\sqrt{3}}\left(\mathbb{Z} + \frac{1}{2}\right)\right)^3 \text{ (s.c.)}$$

$$S_{--} = \left(\frac{2}{\sqrt{3}}\mathbb{Z}\right)^3 \text{ (s.c.)}$$

$$S_{-+} = \left\{\frac{2}{\sqrt{3}}a \,\middle|\, a \in \left(\mathbb{Z} + \frac{1}{2}\right)^3, a_1 + a_2 + a_3 \in 2\mathbb{Z} + \frac{1}{2}\right\} \text{ (f.c.c.).}$$

The zeta-function is

$$Z_{CaF_2}(s) = \frac{1}{2}\left\{Z\left(\frac{4}{3}I, c_0, 0, s\right) - 2Z\left(\frac{4}{3}A, 0, 0, s\right)\right\}$$
$$+ \frac{1}{2}\left\{2Z\left(\frac{4}{3}A, \frac{1}{2}c_0, 0, s\right) - Z\left(\frac{4}{3}I, 0, 0, s\right)\right\}$$
$$= \frac{1}{2}\left(\frac{3}{4}\right)^s Z(I, c_0, 0, s) - \frac{1}{2}\left(\frac{3}{4}\right)^s Z(I, 0, 0, s)$$
$$+ \left(\frac{3}{4}\right)^s Z\left(A, \frac{1}{2}c_0, 0, s\right) - \left(\frac{3}{4}\right)^s Z(A, 0, 0, s),$$

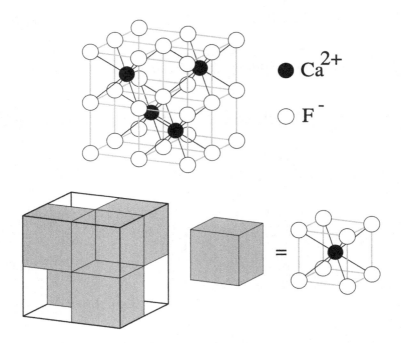

Fig. 6.7   the CaF_2 (Fluorite) structure

for which the following is proved in Example 6.1.

$$Z_{CaF_2}(s) = \frac{1}{2}\left(\frac{3}{4}\right)^s Z_{NaCl}(s) + Z_{CsCl}(s), \qquad (6.67)$$

whence

$$M_{CaF_2} = \frac{\sqrt{3}}{4} M_{NaCl} + M_{CsCl} \qquad (6.68)$$

$$= 2.51939243992\ldots.$$

6.4   Bessel series expansions for Epstein zeta-functions

In this section we shall prove a Bessel series expansion of Chowla-Selberg type (Theorem 6.2) for the Epstein zeta-function $\Lambda(Y, \boldsymbol{g}, \boldsymbol{h}, s)$ corresponding to a block decomposition of the matrix Y. The proof depends on another Bessel series expansion (Theorem 6.1) for the generalized Epstein zeta-function $\sum_{\boldsymbol{a}\in\mathbb{Z}^n} \frac{e^{2\pi i \boldsymbol{h}\cdot\boldsymbol{a}}}{(Y[\boldsymbol{a}+\boldsymbol{g}]+b)^s}$ for $b > 0$, which is interesting in its own right and which we call the Mellin-Barnes type, being dependent on the Mellin-Barnes integrals.

As a corollary to Theorem 6.1, we shall prove the Benson-Mackenzie formula (Corollary 6.1), and for applications of Theorem 6.2, we refer to [KTTY2].

Theorem 6.1 (Mellin-Barnes type formula)   *Notation being as above, we have for $b > 0$,*

$$\pi^{-s} \Gamma(s) \sum_{a \in \mathbb{Z}^n} \frac{e^{2\pi i h \cdot a}}{(Y[a+g]+b)^s}$$

$$= \frac{2}{\sqrt{|Y|}} \sum_{\substack{a \in \mathbb{Z}^n \\ a+h \neq 0}} e^{-2\pi i g \cdot (a+h)} \sqrt{\frac{Y^{-1}[a+h]}{b}}^{s-\frac{n}{2}} K_{s-\frac{n}{2}} \left(2\sqrt{Y^{-1}[a+h]b}\,\pi\right)$$

$$+ \delta(h) \frac{1}{\sqrt{|Y|}} \frac{\Gamma(s-\frac{n}{2})}{\pi^{s-\frac{n}{2}}} \frac{1}{b^{s-\frac{n}{2}}}, \tag{6.69}$$

where $K_s(z)$ signifies the modified Bessel function of the second kind defined by (6.10).

Proof.   This is Formula (1.25) [KTY7] with the term $\varepsilon(g)(\pi b)^{-s}\Gamma(s)$ incorporated in the left-side member. There the proof depended on the modular relation, i.e. the Poisson summation modified so as to suit the case. We refer to Terras for a similar but subtler proof using the Poisson summation formula.

We may deduce (6.69) from the functional equation (6.19) via the Mellin-Barnes integral

$$(1+x)^{-s} = \frac{1}{2\pi i} \int_{(c)} \frac{\Gamma(s-z)\,\Gamma(z)}{\Gamma(s)} x^{-z} \, \mathrm{d}z \tag{6.70}$$

for $x > 0$, $0 < c < \sigma$, which has been used extensively in various context (cf. e.g. [KTZ], [KTTY1], [Matsumoto] and [PK]). The proof starts from expressing the sum in the form of the integral (6.70), applying the functional equation, and then finally appealing to (6.12). $\square$

Corollary 6.1 (Benson-Mackenzie (cf. Borweins' [Bor]))   *Let as before $I = \begin{pmatrix} 1\,0\,0 \\ 0\,1\,0 \\ 0\,0\,1 \end{pmatrix}$, $c_0 = \begin{pmatrix} \frac{1}{2} \\ \frac{1}{2} \\ \frac{1}{2} \end{pmatrix}$ and $I_2 = \begin{pmatrix} 1\,0 \\ 0\,1 \end{pmatrix}$, $c_1 = \begin{pmatrix} \frac{1}{2} \\ \frac{1}{2} \end{pmatrix}$. Then we* *have*

$$Z(I, \mathbf{0}, \mathbf{c}_0, s) = \frac{6\,\pi^{s+1}}{\Gamma(s+1)} \sum_{\mathbf{a} \in (\mathbb{Z} + \frac{1}{2})^2} \sum_{\substack{b \in \mathbb{Z} \\ b \neq 0}} b^2 (-1)^b \sqrt{\frac{I_2[\mathbf{a}]}{b^2}}^{\,s} K_s\left(2\sqrt{I_2[\mathbf{a}]b^2}\,\pi\right)$$

(6.71)

and

$$Z\left(I, \mathbf{0}, \mathbf{c}_0, \frac{1}{2}\right) = -12\pi \sum_{a_1 = \frac{1}{2}}^{\infty} \sum_{a_2 = \frac{1}{2}}^{\infty} \left(\operatorname{sech}\left(\sqrt{a_1^2 + a_2^2}\,\pi\right)\right)^2, \qquad (6.72)$$

Proof.   Since for $\sigma > \frac{3}{2}$

$$Z(I, \mathbf{0}, \mathbf{c}_0, s) = \sum_{\substack{\mathbf{a} \in \mathbb{Z}^3 \\ \mathbf{a} \neq 0}} \frac{(-1)^{a_1 + a_2 + a_3}(a_1^2 + a_2^2 + a_3^2)}{(a_1^2 + a_2^2 + a_3^2)^{s+1}},$$

we may write

$$Z(I, \mathbf{0}, \mathbf{c}_0, s) = 3 \sum_{\substack{b \in \mathbb{Z} \\ b \neq 0}} \sum_{\mathbf{a} \in \mathbb{Z}^2} \frac{(-1)^{a_1 + a_2 + b}\,b^2}{(a_1^2 + a_2^2 + b^2)^{s+1}},$$

whence

$$Z(I, \mathbf{0}, \mathbf{c}_0, s) = 3 \sum_{\substack{b \in \mathbb{Z} \\ b \neq 0}} \left(b^2 (-1)^b \sum_{\mathbf{a} \in \mathbb{Z}^2} \frac{e^{2\pi i \mathbf{c}_1 \cdot \mathbf{a}}}{(I_2[\mathbf{a}] + b^2)^{s+1}} \right). \qquad (6.73)$$

We apply Theorem 6.1 to the inner sum on the right of (6.73) to obtain

$$Z(I, \mathbf{0}, \mathbf{c}_0, s)$$
$$= 3 \sum_{\substack{b \in \mathbb{Z} \\ b \neq 0}} \left(b^2 (-1)^b \frac{2\,\pi^{s+1}}{\Gamma(s+1)} \sum_{\mathbf{a} \in \mathbb{Z}^2} \sqrt{\frac{I_2[\mathbf{a} + \mathbf{c}_1]}{b^2}}^{\,s} K_s\left(2\sqrt{I_2[\mathbf{a} + \mathbf{c}_1]b^2}\,\pi\right) \right),$$

(6.74)

which is (6.71).

Now put $s = \frac{1}{2}$ and recall Formula (6.14) to deduce that

$$Z\left(I, 0, c_0, \frac{1}{2}\right) = 6\pi \sum_{a \in (\mathbb{Z}+\frac{1}{2})^2} \sum_{\substack{b \in \mathbb{Z} \\ b \neq 0}} |b| \, (-1)^b \exp\left(-2\sqrt{I_2[a]} \, |b| \, \pi\right) \qquad (6.75)$$

$$= 12\pi \sum_{a \in (\mathbb{Z}+\frac{1}{2})^2} \sum_{b=1}^{\infty} b \left(-\exp\left(-2\sqrt{I_2[a]} \, \pi\right)\right)^b$$

The inner sum can be evaluated to be

$$\frac{-\exp\left(-2\sqrt{I_2[a]} \, \pi\right)}{\left(1 + \exp\left(-2\sqrt{I_2[a]} \, \pi\right)\right)^2} = -\frac{1}{4}\left(\operatorname{sech}\left(\sqrt{a_1^2 + a_2^2} \, \pi\right)\right)^2,$$

$a = \begin{pmatrix} a_1 \\ a_2 \end{pmatrix} \in \left(\mathbb{Z} + \frac{1}{2}\right)^2$. Hence, splitting the sum over a_1, a_2 into 4 parts, we conclude (6.72), completing the proof. $\qquad\square$

To state Theorem 6.2, we introduce new notation.

Let $Y = \left(\begin{array}{c|c} A & B \\ \hline {}^t B & C \end{array}\right)$ be a block decomposition with A an $n \times n$ matrix and B an $m \times m$ matrix. Set

$$D = C - {}^t B A^{-1} B.$$

In accordance with this decomposition, we decompose the vectors $g = \begin{pmatrix} g_1 \\ g_2 \end{pmatrix}, h = \begin{pmatrix} h_1 \\ h_2 \end{pmatrix}, g_1, h_1 \in \mathbb{Z}^n, g_2, h_2 \in \mathbb{Z}^m$.

Theorem 6.2 (generalized Chowla-Selberg type formula cf. [Ter1, Example 4, p.208]) *Under the above notation, we have*

$$\Lambda(Y, g, h, s) \qquad (6.76)$$

$$= \delta(g_2) \, e^{-2\pi i g_2 \cdot h_2} \Lambda(A, g_1, h_1, s) + \delta(h_1) \frac{1}{\sqrt{|A|}} \Lambda\left(D, g_2, h_2, s - \frac{n}{2}\right)$$

$$+ \frac{2 e^{-2\pi i g_1 \cdot h_1}}{\sqrt{|A|}} \sum_{\substack{a \in \mathbb{Z}^n \\ a+h_1 \neq 0}} \sum_{\substack{b \in \mathbb{Z}^m \\ b+g_2 \neq 0}} e^{2\pi i (-g_1 \cdot a + h_2 \cdot b)} e^{-2\pi i A^{-1} B(b+g_2) \cdot (a+h_1)}$$

$$\times \sqrt{\frac{A^{-1}[a + h_1]}{D[b + g_2]}}^{\, s - \frac{n}{2}} K_{s - \frac{n}{2}}\left(2\sqrt{A^{-1}[a + h_1] \, D[b + g_2]} \, \pi\right).$$

Proof.   The case $g = h = 0, n = m = 1$ is due to Chowla and Selberg [CS], [SC] (cf. also Bateman and Grosswald [BG]), the case $m = 1$ is due to Berndt [Ber6] and the general case with $g = h = 0$ is due to Terras [Ter1].

The proof in our most general case runs as follows.

Noting that

$$\Lambda(Y, g, h, s) = \pi^{-s} \Gamma(s) \sum_{\substack{a \in \mathbb{Z}^n, b \in \mathbb{Z}^m \\ (a+g_1, b+g_2) \neq 0}} \frac{e^{2\pi i(h_1 \cdot a + h_2 \cdot b)}}{Y[(a+g_1, b+g_2)]^s},$$

we distinguish three cases: $b + g_2 = 0$ ($g_2 = 0$ or not) and $b + g_2 \neq 0$:

$$\Lambda(Y, g, h, s) = \delta(g_2) \pi^{-s} \Gamma(s) \sum_{\substack{a \in \mathbb{Z}^n \\ a+g_1 \neq 0}} \frac{e^{2\pi i(h_1 \cdot a - h_2 \cdot g_2)}}{Y[(a+g_1, 0)]^s} \qquad (6.77)$$

$$+ \pi^{-s} \Gamma(s) \sum_{\substack{b \in \mathbb{Z}^m \\ b+g_2 \neq 0}} \sum_{a \in \mathbb{Z}^n} \frac{e^{2\pi i(h_1 \cdot a + h_2 \cdot b)}}{Y[(a+g_1, b+g_2)]^s}.$$

We now apply the formula

$$Y[(a, b)] = A[a + A^{-1}Bb] + D[b], \quad a \in \mathbb{Z}^n, b \in \mathbb{Z}^m \qquad (6.78)$$

to transform (6.77) into

$$\Lambda(Y, g, h, s)$$

$$= \delta(g_2) \pi^{-s} \Gamma(s) e^{-2\pi i h_2 \cdot g_2} \sum_{\substack{a \in \mathbb{Z}^n \\ a+g_1 \neq 0}} \frac{e^{2\pi i h_1 \cdot a}}{A[(a+g_1]^s} \qquad (6.79)$$

$$+ \pi^{-s} \Gamma(s) \sum_{\substack{b \in \mathbb{Z}^m \\ b+g_2 \neq 0}} e^{2\pi i h_2 \cdot b} \sum_{a \in \mathbb{Z}^n} \frac{e^{2\pi i h_1 \cdot a}}{(A[a+g_1+A^{-1}B(b+g_2)]+D[b+g_2])^s}.$$

The first sum on the right of (6.79) is $\Lambda(A, g_1, h_1, s)$ and to the inner sum in the second term, we apply Theorem 6.1. Then the second term on the right of (6.79) becomes

$$
\sum_{\substack{b\in\mathbb{Z}^m, b\neq 0 \\ b+g_2\neq 0}} e^{2\pi i h_2 \cdot b} \left\{ \frac{2}{\sqrt{|A|}} \sum_{\substack{a\in\mathbb{Z}^n \\ a+h_1\neq 0}} e^{-2\pi i (g_1 + A^{-1}B(b+g_2))\cdot(a+h_1)} \right.
$$

$$
\times \left. \sqrt{\frac{A^{-1}[a+h_1]}{D[b+g_2]}}^{\,s-\frac{n}{2}} K_{s-\frac{n}{2}}\left(2\pi\sqrt{A^{-1}[a+h_1]\,D[b+g_2]}\right) \right\}
$$

$$
+ \sum_{\substack{b\in\mathbb{Z}^m \\ b+g_2\neq 0}} e^{2\pi i h_2 \cdot b}\, \delta(h_1) \frac{\Gamma\left(s-\frac{n}{2}\right)}{\sqrt{|A|}\,D[b+g_2]^{\,s-\frac{n}{2}}\,\pi^{s-\frac{n}{2}}},
$$

which are the third and second terms on the right of (6.76), whence the result follows. $\square$

Chapter 7

Fourier series and Fourier transforms

Abstract

This chapter contains elementary facts about Fourier series and transforms; Theorem 7.2 gives sufficient conditions for a Fourier series to converge to the given function $f(t)$, i.e. that $f(t)$ is piecewise of C^1, which is superfluous but sufficient for most of the purpose. In the proof the Fourier series (7.9) for the first periodic Bernoulli polynomial $\overline{B}_1(x)$ is essentially used. Because of its importance, we give two different proofs for (7.9), one by Theorem 7.2, the other by Abel's and Dirichlet's theorem in Appendix B. Regarding integral transforms, we emphasize the case of complex Fourier transforms (and its equivalent form, the Laplace and Mellin transforms). The reader can familiarize oneself with many worked-out concrete examples. This chapter can be read parallel to Chapter 8.

7.1 Fourier series

Suppose f is a periodic function with period $2T$ $(T > 0)$ and integrable over $[-T,\ T]$, and that the procedures made below are all valid.

In analogy with the Laurent expansion (=the Taylor expansion with denominator) $\sum_{n=-\infty}^{\infty} c_n z^n$, we wish to express $f(t)$ in terms of a series in $e^{i\frac{\pi}{T}t}$:

$$\sum_{n=-\infty}^{\infty} c_n \left(e^{i\frac{\pi}{T}t}\right)^n = \sum_{n=-\infty}^{\infty} c_n\, e^{i\frac{n\pi}{T}t}, \qquad (7.1)$$

called the Fourier series $S[f]$ of f.

We want to determine the coefficients c_n. Integrate the series term by term, thereby noting that

$$\frac{1}{2T} \int_{-T}^{T} e^{i\frac{n\pi}{T}t}\, \mathrm{d}t = \begin{cases} 1, & n = 0 \\ 0, & n \neq 0, \end{cases}$$

the orthogonality of the sequence $\{e^{i\frac{n\pi}{T}t}\}_{n\in\mathbb{Z}}$ of complex exponentials, we conclude that

$$\frac{1}{2T} \int_{-T}^{T} f(t)\, e^{-i\frac{n\pi}{T}t}\, \mathrm{d}t = \sum_{m=-\infty}^{\infty} c_m \frac{1}{2T} \int_{-T}^{T} e^{i\frac{(m-n)\pi}{T}t}\, \mathrm{d}t = c_n.$$

Thus the n-th **Fourier coefficient** of f should be defined as

$$c_n = \frac{1}{2T} \int_{-T}^{T} f(t)\, e^{-i\frac{n\pi}{T}t}\, \mathrm{d}t, \tag{7.2}$$

and we may express the **Fourier series** of f which we denote by $S[f]$, as a complex exponential series, in the form

$$S[f] = \sum_{n=-\infty}^{\infty} c_n\, e^{i\frac{n\pi}{T}t}. \tag{7.3}$$

We then write

$$f(t) \sim \sum_{n=-\infty}^{\infty} c_n\, e^{i\frac{n\pi}{T}t}.$$

$$S_n(t) = \sum_{k=-n}^{n} c_k\, e^{i\frac{k\pi}{T}t}, \quad (n \in \mathbb{N}) \tag{7.4}$$

is called the n-th **partial sum**.

Many of the results in the theory of Fourier series may be treated, from a more general standpoint, as those for the orthogonal systems in a vector space. In view of this fact and in anticipation of future progress, let us try to develops a method based on linear algebra. Needless to say, one can come to the same conclusions by direct computation of integrals.

Definition 7.1 The set V of all complex-valued piecewise continuous periodic functions f with period $2T$ forms a complex vector space. In V

we define the **positive semidefinite Hermitian form**

$$(f, g) = \frac{1}{2T} \int_{-T}^{T} f(t) \overline{g(t)} \, dt$$

(cf. Exercise 7.1). (Note that we take the complex conjugation of g). Then V becomes a semi-normed complex vector space with $\|f\| = \sqrt{(f, f)}$ as its semi-norm (length). If $(f, g) = 0$, then we say that f and g are **orthogonal** and write $f \perp g$. A subset $S \subset V$ consisting of vectors of positive length is called an **orthogonal system** if for $\forall f, g \in S$, $f \neq g$, we have $f \perp g$. An orthogonal system with all its vectors having length 1 is called an **orthonormal system** (ONS). Namely, $S \subset V$ is an orthonormal system if and only if

$$(f, g) = \begin{cases} 1, & f = g \\ 0, & f \neq g \end{cases}.$$

An orthonormal system $S \subset V$ is called a **complete** orthonormal system if an addition of one vector having positive length invalidates the orthogonality, or in other words, S is a complete system if and only if an element f of V is orthogonal to all elements of S, then $\|f\| = 0$.

Exercise 7.1 Prove that (f, g) defined above satisfies the conditions (i)—(iv) of the positive semidefinite Hermitian form.

  (i) $(f, f) \geq 0$; $\quad (f, f) = 0 \Leftrightarrow f(t) \equiv 0$ (save for discontinuities)

  (ii) $(f, g) = \overline{(g, f)}$

(iii) $(f + g, h) = (f, h) + (g, h), \quad (f, g + h) = (f, g) + (f, h)$

(iv) $(\lambda f, g) = \lambda (f, g), \quad (f, \lambda g) = \bar{\lambda} (f, g) \quad (\lambda \in \mathbb{C})$

Example 7.1 An example of an orthonormal system.

$$\left\{ \exp\left(i \frac{n\pi}{T} t \right) \right\}_{n \in \mathbb{Z}}$$

is a complete orthonormal system in V. Indeed, we may directly check that

$$\frac{1}{2T} \int_{-T}^{T} \exp\left(i \frac{m\pi}{T} t \right) \overline{\exp\left(i \frac{n\pi}{T} t \right)} \, dt = \frac{1}{2T} \int_{-T}^{T} \exp\left(i \frac{(m - n)}{T} t \right) dt = \delta_{mn}.$$

Lemma 7.1 *If $\{\varphi_1, \ldots, \varphi_n\} \subset V$ ($\varphi_i \neq \varphi_j$ for $i \neq j$) is an orthonormal system, then for any $c_1, \ldots, c_n \in \mathbb{C}$, we have*

$$\left\| \sum_{k=1}^{n} c_k \varphi_k \right\|^2 = \sum_{k=1}^{n} |c_k|^2.$$

Exercise 7.2 Prove Lemma 7.1.

Lemma 7.2 *Let $\{\varphi_1, \ldots, \varphi_n\} \subset V$ ($\varphi_i \neq \varphi_j$ for $i \neq j$) be an orthonormal system and for any $f \in V$, put $c_k = (f, \varphi_k)$. Then we have*

$$\left\| f - \sum_{k=1}^{n} c_k \varphi_k \right\|^2 = \|f\|^2 - \sum_{k=1}^{n} |c_k|^2.$$

Proof. From Exercise 7.1 and Lemma 7.1 it follows that

$$\text{LHS} = \left(f - \sum_{k=1}^{n} c_k \varphi_k, f - \sum_{k=1}^{n} c_k \varphi_k \right)$$

$$= \|f\|^2 - \sum_{k=1}^{n} c_k (\varphi_k, f) - \sum_{k=1}^{n} \bar{c}_k (f, \varphi_k) + \left\| \sum_{k=1}^{n} c_k \varphi_k \right\|^2 = \text{RHS}.$$

$\square$

Corollary 7.1 (Bessel's inequality) *Let $\{\varphi_j\}_{j \in J} \subset V$ ($\varphi_i \neq \varphi_j$ for $i \neq j$) be an orthonormal system and for any $f \in V$, put $c_j = (f, \varphi_j)$. Then $c_j = 0$ except for countably many j and we have*

$$\sum_{j \in J} |c_j|^2 \leq \|f\|^2.$$

In particular, we have $\sum_{j \in J} |c_j|^2 < \infty$.

Remark 7.1 $c_j = (f, \varphi_j)$ *is called the j-th Fourier coefficient of f with respect to the ONS $\{\varphi_j\}_{j \in J}$.*

Corollary 7.2 *If f is a continuous periodic function of period $2T$ and is piecewise of class C^1, then for the Fourier coefficients (7.2) of f, the estimate*

$$\frac{\pi^2}{T^2} \sum_{n=1}^{\infty} n^2 |c_n|^2 \leq \|f'\|^2 < \infty$$

holds.

Proof. f' being piecewise continuous, belongs to V and therefore for its Fourier coefficients

$$\gamma_n = \frac{1}{2T} \int_{-T}^{T} f'(t) e^{-i \frac{n\pi}{T} t} \, dt,$$

we have Bessel's inequality

$$\sum_{n=1}^{\infty} |\gamma_n|^2 \leq \|f'\|^2 . \tag{7.5}$$

Integrating by parts, thereby using the periodicity of f, we get

$$\gamma_n = \frac{1}{2T} \left[f(t) e^{-i\frac{n\pi}{T}t} \right]_{-T}^{T} + \frac{in\pi}{2T^2} \int_{-T}^{T} f(t) e^{-i\frac{n\pi}{T}t} \, dt = i\frac{n\pi}{T} c_n,$$

whence, substituting these in (7.5), we conclude the assertion. $\square$

Exercise 7.3 Compute the right-hand side of

$$\int_{-T}^{T} |f(t) - S_n(t)|^2 \, dt$$

with the aid of Exercise 7.2 and give a direct proof of Bessel's inequality in Corollary 7.1.

Now we shall express (7.3) in a trigonometric form: We write

$$c_n = \frac{1}{2} (a_n - i\, b_n) \quad (n \in \mathbb{Z}),$$

with

$$a_{-n} = a_n, \; b_0 = 0, \; b_{-n} = -b_n \quad (n \in \mathbb{N}),$$

Definition 7.2 Let f be a periodic function of period $2T$ $(T > 0)$ $f(t + 2T) = f(t)$, $t \in \mathbb{R}$. Then we call

$$a_n = \frac{1}{T} \int_{-T}^{T} f(t) \cos \frac{n\pi t}{T} \, dt, \quad b_n = \frac{1}{T} \int_{-T}^{T} f(t) \sin \frac{n\pi t}{T} \, dt, \quad 0 \leq n \in \mathbb{Z} \tag{7.6}$$

the n-th **Fourier cosine coefficient** and **Fourier sine coefficient**, respectively. We have

$$S[f] = \sum_{n=0}^{\infty} A_n(t) = \frac{a_0}{2} + \sum_{n=1}^{\infty} \left(a_n \cos \frac{n\pi t}{T} + b_n \sin \frac{n\pi t}{T} \right) \tag{7.7}$$

and

$$S_n(t) = \sum_{k=0}^{n} A_k(t) = \frac{a_0}{2} + \sum_{k=1}^{n} \left(a_k \cos \frac{k\pi t}{T} + b_k \sin \frac{k\pi t}{T} \right). \tag{7.8}$$

Exercise 7.4 Prove that

$$a_n = c_n + c_{-n}, \quad b_n = i(c_n - c_{-n})$$

and that (7.3) and (7.7) are equivalent.

Exercise 7.5 Prove that if f is a periodic even [odd] function, then $b_n = 0$ [$a_n = 0$] and that

$$S[f] = \frac{a_0}{2} + \sum_{n=1}^{\infty} a_n \cos \frac{n\pi t}{T} \quad \left[S[f] = \sum_{n=1}^{\infty} b_n \sin \frac{n\pi t}{T} \right].$$

This is called the **Fourier cosine** (respectively, the **Fourier sine**) **series**.

Remark 7.2 *Since f is a periodic function of period $2T$, we may choose any interval of length $2T$ as the interval for integration to define the Fourier coefficients. We often use $[-T, T]$ or $[0, 2T]$. Also, by the change of variable $u = \frac{\pi}{T} t$ $\left[u = \frac{t}{2T} \right]$, we may assume the period of f to be 2π [1]. Subsequently, we shall solely consider periodic functions of period $2T$. The Fourier series $S[f]$ for f is just a trigonometrical (exponential) series formed from f and it is not known a priori whether it is convergent or, if convergent, whether it coincides with the original $f(t)$. However, if $f(t)$ is of good-natured, like piecewise smooth (cf. Theorem 7.1), then $S[f](t)$ converges to $f(t)$ at the continuity points of f.*

Exercise 7.6 Define

$$f(t) = t - \frac{1}{2} \quad (0 < t < 1), \quad f(0) = f(1) = 0$$

and extend the domain of definition to all reals by continuing with period 1. Then find the Fourier series of f. Indeed, $f(t) = \overline{B}_1(t)$, $t \notin \mathbb{Z}$ (the first periodic Bernoulli polynomial in Chapter 1).

Solution Since this is the most fundamental, we shall compute the Fourier coefficients a_n, b_n, and c_n. By integration by parts, we have

$$a_n = 2 \int_0^1 \left(t - \frac{1}{2} \right) \cos(2\pi n t) \, dt = 0,$$

$$b_n = 2 \int_0^1 \left(t - \frac{1}{2} \right) \sin(2\pi n t) \, dt = -\frac{1}{\pi n}, \quad n \in \mathbb{N}.$$

Similarly, the n-th Fourier coefficient c_n $(n \neq 0)$ is given by

$$c_n = \int_0^1 \overline{B}_1(x) e^{2\pi i n x} \, dx,$$

which by integration by parts becomes

$$c_n = \left[\frac{1}{2\pi i n} \left(x - \frac{1}{2} \right) e^{2\pi i n x} \right]_0^1 - \frac{1}{2\pi i n x} \int_0^1 e^{2\pi i n x} \, dx = -\frac{1}{2\pi n} i,$$

or

$$a_n = 0, \quad b_n = -\frac{1}{\pi n},$$

as above. Hence

$$f(t) \sim -\frac{1}{\pi} \sum_{n=1}^{\infty} \frac{\sin(2\pi n t)}{n}.$$

Since f is piecewise smooth, we should have the equality (by Theorem 7.1 below)

$$f(t) = -\frac{1}{\pi} \sum_{n=1}^{\infty} \frac{\sin(2\pi n t)}{n}, \quad \forall t \in \mathbb{R}.$$

Hence we have (1.9) in the case of $n = 1$, which we restate for convenience as

$$\overline{B}_1(x) = -\frac{1}{\pi} \sum_{n=1}^{\infty} \frac{\sin 2\pi n x}{n}, \quad x \notin \mathbb{Z}. \tag{7.9}$$

Once (7.9) is established, it is quite easy to deduce Fourier expansion of other linear functions. E.g. consider the function $f(x)$ defined as $x - \pi$ for $0 \leq x < 2\pi$ and continued to a periodic function of period 2π. Since $2\pi \overline{B}_1 \left(\frac{x}{2\pi} \right) = f(x)$, we immediately obtain

$$f(x) = -2 \sum_{n=1}^{\infty} \frac{\sin n x}{n}, \quad x \notin 2\pi \mathbb{Z}. \tag{7.10}$$

Proposition 7.1 (The Riemann-Lebegues Lemma) *Suppose f is piecewise continuous on the interval $[a, b]$. Then all of the following holds true:*

$$\lim_{R \to \infty} \int_a^b f(t) \sin(Rt) \, dt = 0, \quad \lim_{R \to \infty} \int_a^b f(t) \cos(Rt) \, dt = 0,$$

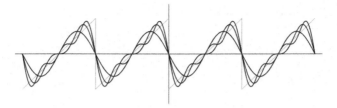

Fig. 7.1

Fig. 7.2

Fig. 7.3

$$\lim_{R\to\infty} \int_a^b f(t)\, e^{iRt}\, \mathrm{d}t = 0.$$

Proof. We may assume that f is continued on the whole interval $[a, b]$
(Exercise 7.7). Then, being continuous on the compact set $[a, b]$, f is
bounded (Weierstrass' Theorem), i.e. $f(t) = O(1)$, $t \in [a, b]$ and uniformly
continuous, whence

$$f\left(u + \frac{\pi}{R}\right) - f(u) = o(1), \quad R \to \infty.$$

We shall prove the first assertion. Putting $t = u + \frac{\pi}{R}$, then

$$I := \int_a^b f(t) \sin(Rt) \, dt = - \int_{a-\pi/R}^{b-\pi/R} f\left(u + \frac{\pi}{R}\right) \sin(Ru) \, du.$$

Hence

$$2I = I - \int_{a-\pi/R}^{b-\pi/R} f\left(u + \frac{\pi}{R}\right) \sin(Ru) \, du$$

$$= \int_{b-\pi/R}^b f(u) \sin(Ru) \, du + \int_a^{b-\pi/R} \left(f(u) - f\left(u + \frac{\pi}{R}\right)\right) \sin(Ru) \, du$$

$$- \int_{a-\pi/R}^a f\left(u + \frac{\pi}{R}\right) \sin(Ru) \, du$$

$$= O\left(\frac{1}{R}\right) + o(1) + O\left(\frac{1}{R}\right) = o(1), \quad R \to \infty.$$

$\square$

Exercise 7.7 (i) Divide $[a, b]$ into subintervals to prove that the above proof can be reduced to the case where $f(t)$ is continuous on the whole interval. (ii) Prove the remaining assertions of Proposition 7.1.

Theorem 7.1 *If f is a periodic function of period $2T$ and is piecewise smooth as well as continuous on any finite interval, then its Fourier series (7.3) converges to $f(t)$ uniformly on any finite interval.*

Proof. First we show that $\{S_n\}$ is convergent, where $S_n = S_n(t)$ is the n-th partial sum defined by (7.4):

For integers $N > M > 0$, we have

$$|S_N - S_M| = \left| \sum_{|k|=M+1}^N k \, c_k \frac{1}{k} e^{i\frac{k\pi}{T}t} \right|$$

$$\leq \sqrt{\sum_{|k|=M+1}^N k^2 |c_k|^2} \sqrt{\sum_{|k|=M+1}^N \frac{1}{k^2} \left| e^{i\frac{k\pi}{T}t} \right|^2}$$

$$\leq \frac{T}{\pi} \|f'\| \sqrt{\sum_{k=M+1}^N \frac{1}{k^2}}$$

by the Canchy-Schwarz inequality and the Bessel inequality (Corollary 7.1). Hence $|S_N - S_M| \to 0$ as $N, M \to \infty$, and the Canchy criterion applies, ensuring the convergence of $\{S_n\}$. We now show that its limit is $f(t)$.

Recalling (7.3), we may express (7.4) as

$$S_n = \sum_{k=-n}^{n} \frac{1}{2T} \int_{-T}^{T} e^{-i\frac{k\pi}{T}u} e^{i\frac{k\pi}{T}t} f(u)\,du$$

$$= \frac{1}{T} \int_{-T}^{T} \frac{1}{2} \sum_{k=-n}^{n} e^{i\frac{k\pi}{T}(t-u)} f(u)\,du.$$

Now the **Dirichlet kernel** is

$$\frac{1}{2} + \sum_{k=1}^{n} \cos \frac{k\pi}{T}(t-u),$$

which by Exercise 7.8, (2), $\dfrac{\sin\left(n+\frac{1}{2}\right)\frac{\pi}{T}(t-u)}{2\sin\frac{\pi}{T}\frac{t-u}{2}}.$

Hence

$$S_n(t) = \frac{1}{T} \int_{-T}^{T} \frac{\sin\left(n+\frac{1}{2}\right)\frac{\pi}{T}(t-u)}{2\sin\frac{\pi}{T}\frac{t-u}{2}} f(u)\,du$$

$$= \frac{1}{T} \int_{-T-t}^{T-t} \frac{\sin\left(n+\frac{1}{2}\right)\frac{\pi}{T}v}{2\sin\frac{\pi}{T}\frac{v}{2}} f(t+v)\,dv$$

$$= \frac{1}{T} \int_{-T}^{T} \frac{\sin\left(n+\frac{1}{2}\right)\frac{\pi}{T}u}{2\sin\frac{\pi}{T}\frac{u}{2}} f(t+u)\,du \qquad (7.11)$$

by change of variable and the periodicity of f.

Subtracting $f(t) = \frac{1}{T} \int_{-T}^{T} \frac{\sin\left(n+\frac{1}{2}\right)\frac{\pi}{T}u}{2\sin\frac{\pi}{T}\frac{u}{2}} f(t)\,du$ from (7.11) and using

$$\frac{\sin\left(n+\frac{1}{2}\right)\frac{\pi}{T}u}{\sin\frac{\pi}{T}\frac{u}{2}} = \frac{\sin n\frac{\pi}{T}u \cos\frac{\pi}{T}\frac{u}{2}}{\sin\frac{\pi}{T}\frac{u}{2}} + \cos n\frac{\pi}{T}u,$$

we find that

$$S_n(t) - f(t) = \frac{1}{2T} \int_{-T}^{T} g(u,t) \sin n\frac{\pi}{T}u\,du$$

$$+ \frac{1}{2T} \int_{-T}^{T} (f(u+t) - f(u)) \cos n\frac{\pi}{T}u\,du,$$

where

$$g(u,t) = \frac{(f(u+t) - f(u)) \cos\frac{\pi}{T}\frac{u}{2}}{\sin\frac{\pi}{T}\frac{u}{2}}$$

whose possible discontinuity, save for those of f, is at $u = 0$.

However, by the piecewise continuity of $f'(t)$,

$$g(u,t) = \frac{f(u+t) - f(u)}{u} \frac{\frac{\pi}{T}\frac{u}{2}\cos\frac{\pi}{T}\frac{u}{2}}{\sin\frac{\pi}{T}\frac{u}{2}} \frac{2T}{\pi}$$

$$\rightarrow \begin{cases} \dfrac{2T}{\pi} f'(t+), & \text{as } u \to 0+ \\[2mm] \dfrac{2T}{\pi} f'(t-), & \text{as } u \to 0 - . \end{cases}$$

Hence $g(u,t)$ is also piecewise continuous and the Riemann-Lebesgue Lemma (Proposition 7.1) shows that

$$\lim_{n\to\infty} S_n(t) = f(t).$$

$\square$

Theorem 7.2 *If f is periodic of period $2T$, piecewise continuous and piecewise of C^1, then the Fourier series for $f(t)$ converges to $\frac{1}{2}(f(t+) + f(t-))$.*

Proof. We may consider the case where 0 is the only discontinuity of f, other cases being reduced to this. Then the function

$$F(t) = f(t) + (f(0+) - f(0-))\, g(t)$$

is piecewise of C^1 and continuous except possibly at 0, where

$$g(t) = \begin{cases} \overline{B}_1\left(\frac{t}{2T}\right), & t \notin 2T\mathbb{Z} \\ 0, & t \in 2T\mathbb{Z}. \end{cases}$$

But

$$\lim_{t\to 0-} F(t) = f(0-) + \frac{1}{2}(f(0+) - f(0-)) = \frac{1}{2}(f(0+) + f(0-))$$

and

$$\lim_{t\to 0+} F(t) = f(0+) - \frac{1}{2}(f(0+) - f(0-)) = \frac{1}{2}(f(0+) - f(0-)),$$

and therefore $F(t)$ is also continuous at 0.

Hence we may apply Theorem 7.1 to conclude that F has the Fourier series which converges to $F(t)$ everywhere. But $g(t)$ has the Fourier series

$$-\frac{1}{\pi} \sum_{n=1}^{\infty} \frac{\sin\frac{\pi}{T}nt}{n}$$

converging to $g(t)$ everywhere (cf. Exercise 7.6).

Hence it follows that $f(t)$ also has the Fourier series converging to it at continuities and to $F(0) = \frac{1}{2}(f(0+) - f(0-))$, thereby completing the proof. $\square$

Exercise 7.8   For $x \notin \mathbb{Z}$ prove that

$$\sum_{k=1}^{n} e^{2\pi i k x} = e^{\pi i (n+1)x} \frac{\sin \pi n x}{\sin \pi x} \qquad (7.12)$$

and deduce from (7.12) that

$$\sum_{k=1}^{n} \cos 2\pi k x + \frac{1}{2} = \frac{\sin(2n+1)\pi x}{2 \sin \pi x} \qquad (7.13)$$

and

$$\sum_{k=1}^{n} \sin 2\pi k x = \sin((n+1)\pi x) \frac{\sin \pi n x}{\sin \pi x} = \frac{\cos \pi x - \cos \pi(2n+1)x}{2 \sin \pi x} \qquad (7.14)$$

Solution   We have

$$S_n = \sum_{k=1}^{n} e^{2\pi i k x} = \frac{e^{2\pi i x}\left(e^{2\pi i n x} - 1\right)}{e^{\pi i x} - 1} = \frac{e^{2\pi i n x} - 1}{1 - e^{-2\pi i x}}.$$

We factor out $e^{\pi i n x}$ (resp. $e^{-\pi i x}$) from the numerator (resp. denominator) to get

$$S_n = e^{\pi i (n+1)x} \frac{\frac{e^{\pi i n x} - e^{-\pi i n x}}{2i}}{\frac{e^{\pi i x} - e^{-\pi i x}}{2i}},$$

which is (7.12). The real part of (7.12) is

$$\sum_{k=1}^{n} \cos 2\pi k x = \frac{\cos(n+1)\pi x \sin \pi n x}{\sin \pi x} = \frac{\sin \pi(2n+1)x - \sin \pi x}{2 \sin \pi x},$$

which proves (7.13). The imaginary part of (7.12) is the same as (7.14).
    Another proof uses the formulae

$$\cos \alpha \sin \beta = \frac{1}{2}\left(\sin(\alpha + \beta) - \sin(\alpha - \beta)\right)$$

and

$$\sin \alpha \sin \beta = -\frac{1}{2}\left(\cos(\alpha + \beta) - \cos(\alpha - \beta)\right).$$

The first gives

$$\sum_{k=1}^{n} \cos 2\pi kx \sin \pi x$$

$$= \frac{1}{2} \sum_{k=1}^{n} (\sin(2k+1)\pi x - \sin(2k-1)\pi x)$$

$$= \frac{1}{2} \big(\sin(2n+1)\pi x - \sin \pi x \big),$$

or

$$\sum_{k=1}^{n} \cos 2\pi kx = \frac{1}{2} \frac{\sin(2n+1)\pi x}{\sin \pi x} - \frac{1}{2},$$

which is (7.13).

The second gives

$$\sum_{k=1}^{n} \sin 2\pi kx \sin \pi x$$

$$= -\frac{1}{2} \sum_{k=1}^{n} (\cos(2k+1)\pi x - \cos(2k-1)\pi x)$$

$$= -\frac{1}{2} \big(\cos(2n+1)\pi x - \cos \pi x \big),$$

which is (7.14).

Example 7.2 Let $f(x)$ be defined for $-1 \leq x < 1$ as

$$f(x) = \begin{cases} 0, & -1 \leq x < 0 \\ 1, & 0 \leq x < 1, \end{cases}$$

and then defined periodically with period 2:

$$\overline{f}(x) = f(x + 2n), \ n \in \mathbb{Z}.$$

(cf. Fig. 7.4 for its graph).

This function can be represented as

$$\overline{f}(x) = [x+1] - 2 \left[\frac{x+1}{2} \right]. \tag{7.15}$$

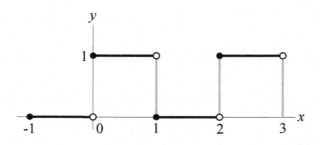

Fig. 7.4

Recalling $\overline{B}_1(x) = x - [x] - \dfrac{1}{2}$, we have $[x] = x - \overline{B}_1(x) - \dfrac{1}{2}$, so that

$$\overline{f}(x) = x + 1 - \overline{B}_1(x+1) - \frac{1}{2} - 2\left(\frac{x+1}{2} - \overline{B}_1\left(\frac{x+1}{2} - \frac{1}{2}\right)\right)$$

$$= \frac{1}{2} + 2\overline{B}_1\left(\frac{x+1}{2}\right) - \overline{B}_1(x+1).$$

Applying the Fourier expansion (7.9) for $\overline{B}_1(x)$ below, we see that

$$\overline{f}(x) = \frac{1}{2} - \frac{2}{\pi}\sum_{n=1}^{\infty}\frac{\sin 2\pi n \frac{x+1}{2}}{n} + \frac{1}{\pi}\sum_{n=1}^{\infty}\frac{\sin 2\pi n x}{n}$$

$$= \frac{1}{2} + \frac{1}{\pi}\sum_{n=1}^{\infty}\frac{2}{n}(-1)^{n-1}\sin\pi n x + \frac{1}{\pi}\sum_{n=1}^{\infty}\frac{\sin 2\pi n x}{n}$$

$$= \frac{1}{2} + \frac{2}{\pi}\sum_{m=1}^{\infty}\frac{2}{2m}(-1)^{2m-1}\sin 2\pi m x$$

$$+ \frac{1}{\pi}\sum_{m=1}^{\infty}\frac{2}{2m-1}(-1)^{2m-2}\sin(2m-1)\pi x + \frac{1}{\pi}\sum_{n=1}^{\infty}\frac{1}{n}\sin 2\pi n x$$

$$= \frac{1}{2} + \frac{2}{\pi}\sum_{m=1}^{\infty}\frac{1}{2m-1}\sin(2m-1)\pi x, \quad x \notin \mathbb{Z}.$$

For $x \in \mathbb{Z}$, the Fourier series converges to 0. To deduce (7.15) we argue in the following manner. First $y = [x+1]$ has the graph (Figure 7.8):

For $2n - 1 \le x < 2n + 1$, we have to pull down this graph by $2n$, i.e. by $2\left[\dfrac{x+1}{2}\right]$ since $n \le \dfrac{x+1}{2} < n+1$ implies $\left[\dfrac{x+1}{2}\right] = n$.

Example 7.3   There is a method much subtler and more ingenious than

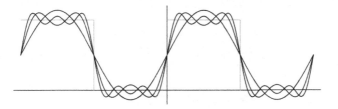

Fig. 7.5

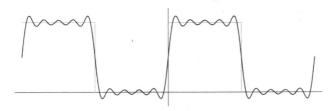

Fig. 7.6

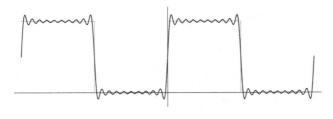

Fig. 7.7

the customary one for deducing (7.9), which appeals to Abel's continuity theorem.

We wish to apply this theorem to the Maclaurin expansion for

$$-\log(1 - z).$$

The expansion can most easily be obtained by termwise integration of the absolutely convergent power series

$$\sum_{n=0}^{\infty} z^n = \frac{1}{1 - z}, \quad |z| < 1. \tag{7.16}$$

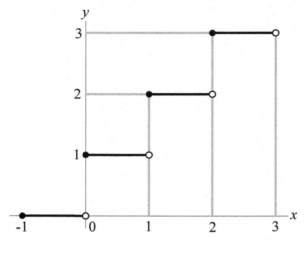

Fig. 7.8

Indeed, integrating from 0 to z, we obtain

$$-\log(1-z) = \int_0^z \frac{1}{1-z}\,\mathrm{d}z = \sum_{n=0}^{\infty} \int_0^z z^n\,\mathrm{d}z = \sum_{n=1}^{\infty} \frac{1}{n}\,z^n,$$

where we have taken the principal branch of the logarithm.

Now, for $z = e^{i\theta}$, $\theta \notin 2\pi\mathbb{Z}$, we contend that the series $\sum_{n=1}^{\infty} \frac{1}{n} e^{in\theta}$ is convergent, as shown in Example B.2. Hence, Abel's theorem B.6 allows us to write

$$-\log\left(1 - e^{i\theta}\right) = \sum_{n=1}^{\infty} \frac{1}{n}\left(e^{i\theta}\right)^n, \qquad \theta \notin 2\pi\mathbb{Z}. \tag{7.17}$$

By Euler's identity, the right-hand side is

$$\sum_{n=1}^{\infty} \frac{1}{n}\cos n\theta + i \sum_{n=1}^{\infty} \frac{1}{n}\sin n\theta,$$

while the left-hand side may be transformed into

$$- \log \left(1 - e^{i\theta}\right) = - \log \frac{e^{i\frac{\theta}{2}} - e^{-i\frac{\theta}{2}}}{2i} \cdot \frac{-2i}{e^{-i\frac{\theta}{2}}}$$

$$= - \log \sin \frac{\theta}{2} - \log(-2i) + \log e^{-\frac{i\theta}{2}}$$

$$= - \log 2 \sin \frac{\theta}{2} - \frac{i}{2} (\theta - \pi),$$

since $\log(-2i) = \log 2 - \frac{\pi}{2} i$. Hence, comparing the real and imaginary parts, we conclude that

$$- \log 2 \sin \frac{\theta}{2} = \sum_{n=1}^{\infty} \frac{1}{n} \cos n\theta \qquad (7.18)$$

$$\theta - \pi = -2 \sum_{n=1}^{\infty} \frac{1}{n} \sin n\theta. \qquad (7.19)$$

Thus we have not only recovered (7.10) again but obtained the Fourier expansion of the log sin function.

Example 7.4 The function $f(x) = \begin{cases} 0, & -1 \le x < 0 \\ x, & 0 \le x < 1 \end{cases}$ can be expressed as $\frac{1}{2}(x + |x|)$, the positive part $f^+(x)$ of $f(x) = x$, where the positive part of $f(x)$ is defined to be

$$f^+(x) = \frac{f(x) + |f(x)|}{2}.$$

The function $\overline{f}(x)$ (for its graph, see Fig. 7.9) obtained from $f(x)$ by continuing it periodically with period 2 is

$$\overline{f}(x) = \frac{1}{2} \left(x - 2 \left[\frac{x+1}{2} \right] + \left| x - 2 \left[\frac{x+1}{2} \right] \right| \right). \qquad (7.20)$$

Indeed, for $2n - 1 \le x < 2n + 1$ ($n \in \mathbb{Z}$), we have $n \le \frac{x+1}{2} < n + 1$, and so $\left[\frac{x+1}{2} \right] = n$. Hence for $2n - 1 \le x < 2n$, we have $x - 2 \left[\frac{x+1}{2} \right] = x - 2n < 0$, so that $\overline{f}(x) = \frac{1}{2}(x - 2n + 2n - x) = 0$. For $2n \le x < 2n + 1$, we have $x - 2 \left[\frac{x+1}{2} \right] = x - 2n \ge 0$, and therefore $\overline{f}(x) = 2(x - 2n) = x - |x|$.

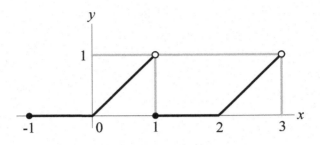

Fig. 7.9

Noting that $2\left[\frac{x+1}{2}\right] = x - 2\overline{B}_1\left(\frac{x+1}{2}\right)$, we deduce that

$$\overline{f}(x) = \frac{1}{2}\left(2\overline{B}_1\left(\frac{x+1}{2}\right) + \left|2\overline{B}_1\left(\frac{x+1}{2}\right)\right|\right) \qquad (7.21)$$

$$= \overline{B}_1\left(\frac{x+1}{2}\right) + \left|\overline{B}_1\left(\frac{x+1}{2}\right)\right|$$

We remark that the graph of $y = \left|\overline{B}_1\left(\dfrac{x+1}{2}\right)\right|$ is the directly connected infinite tents (see Fig. 7.10).

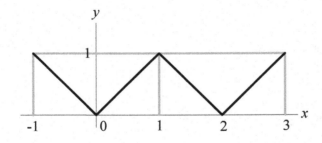

Fig. 7.10

Of course, if we recall this fact first, then the expression (7.21) would follow by inspection.

We also note that if we replace the period 2 by 2π, we get

$$\overline{f}_1(x) = \frac{1}{2}\left(\frac{1}{\pi}x - 2\left[\frac{x+\frac{1}{2}}{2\pi}\right] + \left|\frac{1}{\pi}x - 2\left[\frac{x+\frac{1}{2}}{2\pi}\right]\right|\right). \qquad (7.22)$$

Motivated by the above function, we shall digress here into an equivalent statement of the celebrated Riemann Hypothesis (abbreviated as RH) to the effect that the Riemann zeta-function does not vanish on the central line $\sigma = \frac{1}{2}$.

We need some preparations.

Let $t(x)$ denote the tent function

$$
t(x) = \begin{cases} 2x, & 0 \leq x \leq \frac{1}{2} \\ 2 - 2x, & \frac{1}{2} \leq x \leq 1. \end{cases}
$$

We consider the $(k+1)$-th iterate of t divided by 2^{k+1}:

$$
f_{2^k}(x) = \frac{1}{2^{k+1}} t^{k+1}(x),
$$

where by an iterate we mean the successive composition of t:

$$
t^k(x) = t\big(t^{k-1}(x)\big), \quad t^1(x) = t(x), \quad t^0(x) = 1.
$$

Denoting the directly connected n tents by $f_n(x)$ with length $\frac{1}{n}$ and height $\frac{1}{2n}$, we note that $f_{2^k}(x) = t^k(x)$ in the case $n = 2^k$.

The Farey sequence F_x of order $[x]$ is defined to be the increasing sequence of irreducible fractions ρ_ν between 0 and 1 (0 exclusive) with denominators $\leq x$. This may be constructed from the lower order one by inserting the mediants of adjacent fractions until the denominator reaches $[x]$. The total number of elements of F_x is $\sum_{n \leq x} \phi(n) = \Phi(x)$, say, where $\varphi(n)$ indicates the Euler function (cf. (8.27)).

For any even integrable core function f on $[0,1]$ we define the error term

$$
E_f(x) = \sum_{\nu=1}^{\Phi(x)} f(\rho_\nu) - \Phi(x) \int_0^1 f(x)\,\mathrm{d}x,
$$

where by an even function we mean that it satisfies $f(x) = f(1-x)$, which we may assume on symmetry grounds. $E_f(x)$ is defined to be 0 for $0 < x < 1$.

We consider the Mellin transform (cf. §7.4) of $E_f(x)$:

$$
F(s) = s\,\zeta(s) \int_0^\infty E_f(x)\, x^{-s}\, \frac{\mathrm{d}x}{x}.
$$

Then we have

Lemma 7.3 ([BKY, Lemmas 2.1 and 2.2]) *(i) For the directly connected n tents $f_n(x)$, the associated Mellin transform $F(s)$ is given by*

$$F(s) = \frac{1}{12n} F_n(s),$$

where

$$F_n(s) = \sum_{m=1}^{\infty} \frac{c_n(m)}{m^{s+1}}$$

and

$$c_n(m) = (m, 2n)^2 - (2m, 2n)^2.$$

This $F_n(s)$ can be written down as follows:

$$F_n(s) = -3\left(1 - \frac{1}{2^{s+1}}\right) \zeta(s+1) C_n(s),$$

where

$$C_n(s) = \sum_{d|n} d^{1-s} \sum_{\substack{\delta|d \\ d:\text{odd}}} \frac{\mu(\delta)}{\delta^2}.$$

(ii) For $n = 2^k$ we have

$$C_{2^k}(s) = \frac{1 - 2^{(1-s)(1+k)}}{1 - 2^{1-s}}$$

which does not vanish for $\frac{1}{2} < \delta < 1$.

Theorem 7.3 ([BKY, Theorem 2.1 (i)]) *The asymptotic formula*

$$\sum_{\nu=1}^{\Phi(x)} f_{2^k}(\rho_\nu) = \frac{1}{2^k} \Phi(x) + O\left(x^{\frac{1}{2}+\epsilon}\right),$$

for every $\epsilon > 0$, is equivalent to the Riemann Hypothesis.

Example 7.5 The function defined for $-\pi \le x < \pi$ by $f(x) = x$ and continued to be a periodic function with period 2π is

$$\overline{f}(x) = \pi \overline{B}_1\left(\frac{1}{2\pi}x + \frac{1}{2}\right).$$

Fig. 7.11   laplace

7.2   Integral transforms

Irrespectively of whether it is in pure or applied areas, a great majority of important special functions that appear in applications of analysis are given in the form of an improper integral of a real function

$$(Kf)(s) = \int_{-\infty}^{\infty} K(x,s)\,f(x)\,\mathrm{d}x$$

$$= \left(\lim_{\lambda_1 \to -\infty} \int_{\lambda_1}^{t} + \lim_{\lambda_2 \to \infty} \int_{t}^{\lambda_2} \right) K(x,s)\,f(x)\,\mathrm{d}x,$$

where $s = \sigma + it$ signifies the complex variable. We call $(Kf)(s)$ an integral transform of f (with respect to the kernel function K) in view of the fact that it is obtained by integrating after multiplying f by the (complex-valued) kernel function $K(x, s)$. In contrast to differential operators D, Δ, ∇ that we have already learned, K, being an operator exerting the integration of f, is called an integral operator. The integral transform of f is denoted by the corresponding capital letter F and is often denoted by $F_K(s)$ in order to suggest the (commonly accepted name of) the kernel:

$$(Kf)(s) = F_K(s) = \int_{-\infty}^{\infty} K(x, s)\, f(x)\, dx.$$

We refer to $F_K(s)$ as the resulting integral transform of f. When we know which integral transform, we may suppress K and simply write $F(s)$. Integral transforms and operators are useful in solving Boundary Value Problems and Differentiable [integral] Equations. However, we have to impose stringent conditions for the improper integrals to converge. Since the integrals being intrinsically linear, so are the improper ones in the region of their convergence, and a fortiori so are the integral operators. Namely, for c_1, $c_2 \in \mathbb{C}$,

$$K(c_1 f_1 + c_2 f_2) = c_1 K f_1 + c_2 K f_2.$$

The following integral transforms are most frequently used.

$$(Ff)(s) = \hat{f}(s) = \int_{-\infty}^{\infty} e^{-ixs} f(x)\, dx \quad \text{(exponential) Fourier transform}$$

$$(Cf)(s) = F_C(s) = \int_{-\infty}^{\infty} \cos(xs)\, f(x)\, dx \quad \text{Fourier (cosine) transform}$$

$$(Sf)(s) = F_S(s) = \int_{-\infty}^{\infty} \sin(xs)\, f(x)\, dx \quad \text{Fourier (sine) transform}$$

$$(L_I f)(s) = F_{LI}(s) = \int_{0}^{\infty} e^{-xs} f(x)\, dx \quad \text{(one-sided) Laplace transform}$$

$$(L_{II} f)(s) = F_{LII}(s) = \int_{-\infty}^{\infty} e^{-xs} f(x)\, dx \quad \text{(two-sided) Laplace transform}$$

$$(Mf)(s) = F_M(s) = \int_{0}^{\infty} x^{s-1}\, f(x)\, dx \quad \text{Mellin transform}$$

We may express a function that describes the state of a phenomenon at some time interval as a real function $f(t)$ in the time variable t. We call a function a **causal function** if it is not affected by the conditions before the initial time $t = 0$, i. e. if $f(t) = 0$, $t < 0$. It follows from Euler's identity that

$$(Cf)(s) = \frac{1}{2}(Ff)(s) + \frac{1}{2}(Ff)(-s), \quad (Sf)(s) = \frac{1}{2i}(Ff)(s) - \frac{1}{2i}(Ff)(-s),$$

whence we see that both the cosine and sine Fourier transforms are special cases of the exponential Fourier transforms.

The region of convergence of the exponential Fourier transform (if it exists) must be a strip $\gamma_1 < \operatorname{Im} s < \gamma_2$ containing the real axis (because t takes values both positive and negative). However, whenever we speak of a Fourier transform, we mean the case $s = y$ being a real variable, we shall refer to this as a **Fourier transform**

$$(Ff)(y) = \hat{f}(y) = \int_{-\infty}^{\infty} e^{-ixy} f(x)\,dx, \quad y \in \mathbb{R},$$

and the exponential Fourier transform as the **complex Fourier transform**. Then the two-sided Laplace transform is no other than the general Fourier transform

$$(L_{II}f)(is) = (Ff)(s)$$

in the complex variable s, rotated clockwise by $90°$, and so their theory is almost parallel and can be translated word-for-word into each other. E. g. the region of convergence (for two-sided Laplace transform) is a strip $\beta_1 < \operatorname{Re} s < \beta_2$ containing the imaginary axis (this is analogous to the fact stated in § A.1 that if $f(z)$ is analytic on the unit circle $|z| = 1$, then it can be expanded into the Laurent series and

$$g(\theta) = f(e^{i\theta})$$

can be expressed as a Fourier series. If the annulus does not contain the unit circle, then we cannot write $g(\theta) = f(e^{i\theta})$ even if $g(\theta)$ can be expanded into Fourier series). There is a theory of operators developed on the basis of two-sided Laplace transforms (cf. [Pa]). It looks as if the Laplace transforms have driven out Fourier transforms in applied analysis (on the ground that the former seem to have a wider applicability than the latter), but they are essentially the same. In comparison with the real Fourier transforms, the condition that the improper integral for $(L_{II}f)(s)$ be convergent in the strip $\beta_1 < \operatorname{Re} s < \beta_2$ restricts the class of functions than the condition that the improper integral for $(Ff)(y)$ be convergent for any $y \in \mathbb{R}$.

Indeed, putting $s = \sigma + it$ and $\varphi_\sigma(x) = e^{-x\sigma} f(x)$, then we have

$$(L_{II}f)(s) = (L_{II}f)(\sigma + it) = \int_{-\infty}^{\infty} e^{-ixt} \varphi_\sigma(tx)\, dx = (F\varphi_\sigma)(t),$$

σ being regarded as a parameter, so that "the Laplace transform, the range of whose variable is restricted to a vertical line in its region of convergence, is a special case of the Fourier transform and vice versa". More restrictive it is, the merit of Laplace transforms is that the variable ranging in a strip without restriction, we may employ the powerful theory of analytic functions. For a causal function, its two-sided and one-sided Laplace transforms are the same and the region of convergence (if it exists) is the half-plane $\operatorname{Re} s = \sigma > \beta$ with a wide range of applications. In what follows, for a causal function $f(x)$ $(f(x) = 0,\ x < 0)$, we shall call its one-sided Laplace transform simply the **Laplace transform**;

$$(Lf)(s) = F(s) = \int_0^{\infty} e^{-xs} f(x)\, dx, \quad s \in \mathbb{C}.$$

In electrical engineering, with ω being the angular cycle, the variable is often denoted by $s = \sigma + j\omega$ (j being the imaginary unit) or $p = \sigma + j\omega$. Use being made of the Fourier transform $\hat{f}(y)$, the original function $f(x)$ which behaves differently on different parts of the x-axis may be expressed by a unique formula

$$f(x) = \frac{1}{2\pi} \int_{-\infty}^{\infty} \hat{f}(y)\, e^{iyx}\, dy. \tag{7.23}$$

This is called the **Fourier Integral Theorem**. To be more precise, if f, f' are piecewise continuous and

$$\int_{-\infty}^{\infty} |f(x)|\, dx < \infty,$$

then the theorem holds in the following form:

$$\frac{1}{2}\{f(x+0) + f(x-0)\} = \frac{1}{2\pi} \int_{-\infty}^{\infty} \hat{f}(y)\, e^{iyx}\, dy. \tag{7.24}$$

If we define the **inverse Fourier transform** $F^{-1}f$ of f by

$$\left(F^{-1}f\right)(y) = \frac{1}{2\pi} \int_{-\infty}^{\infty} f(x)\, e^{ixy}\, dx, \tag{7.25}$$

then we may express the Fourier Integral Theorem as $F^{-1}Ff = f$ or $\widehat{\hat{f}}(-x) = f(x)$. (7.23) may be deduced formally from Theorem 7.2 as follows. If f is piecewise smooth and continuous on $[-T, T]$, then it can be expanded into a Fourier series:

$$f(t) = \sum_{n=-\infty}^{\infty} c_n\, e^{i\lambda_n t}, \quad c_n = \frac{1}{T} \int_{-T/2}^{T/2} f(x)\, e^{-i\lambda_n x}\, dx,$$

$$\lambda_n = \frac{2\pi}{T} n, \quad |t| < \frac{T}{2}.$$

Letting $T \to \infty$, $n \to \infty$, we may contend that

$$T c_n \to \int_{-\infty}^{\infty} f(x)\, e^{-i\lambda_n x}\, dx = \hat{f}(\lambda_n)$$

and therefore

$$f(t) = \frac{1}{2\pi} \sum_{n=-\infty}^{\infty} T c_n e^{i\lambda_n t} (\lambda_{n+1} - \lambda_n)$$

$$\to \frac{1}{2\pi} \sum_{n=-\infty}^{\infty} \hat{f}(\lambda_n)\, e^{i\lambda_n t} \Delta\lambda_n \sim \frac{1}{2\pi} \int_{-\infty}^{\infty} \hat{f}(y)\, e^{iyt}\, dy.$$

Viewing this as

$$f(t) = \int_{-\infty}^{\infty} \left(\frac{1}{2\pi} \int_{-\infty}^{\infty} f(x)\, e^{-ixy}\, dx \right) e^{iyt}\, dy,$$

this may be thought of as giving the motivation for the definition of $\hat{f}(x)$. Also using the defining equation

$$\int_{-\infty}^{\infty} f(x)\, \delta(t - x)\, dx = f(t)$$

for the delta function and one of its well known properties

$$\frac{1}{2\pi} \int_{-\infty}^{\infty} e^{ixt}\, dx = \delta(t),$$

we can give the following simple proof.

$$\frac{1}{2\pi} \int_{-\infty}^{\infty} (Ff)(y)\, e^{iyt}\, dy = \int_{-\infty}^{\infty} \left(\frac{1}{2\pi} \int_{-\infty}^{\infty} e^{-ixy} f(x)\, dx \right) e^{iyt}\, dy$$

$$= \int_{-\infty}^{\infty} f(x)\, dx \left(\frac{1}{2\pi} \int_{-\infty}^{\infty} e^{iy(t-x)}\, dy \right)$$

$$= \int_{-\infty}^{\infty} f(x)\, \delta(t-x)\, dx = f(t).$$

This is a legitimate proof if the inversion of the order of integration is justified. A rigorous proof can be given in the spirit of Proof of Theorem 7.2, which is omitted. In view of the appearance of the factor $\frac{1}{2\pi}$ in the Fourier integral theorem, we often introduce normalization by distributing it to both transforms. The symmetric **pair of the Fourier transform** and the inverse Fourier transform is

$$(Ff)(y) = \hat{f}(y) = \frac{1}{\sqrt{2\pi}} \int_{-\infty}^{\infty} e^{-ixy} f(x)\, dx,$$

$$(F^{-1}f)(y) = \frac{1}{\sqrt{2\pi}} \int_{-\infty}^{\infty} f(x)\, e^{ixy}\, dx.$$

Then the Fourier cosine transform takes the form for f even,

$$(Cf)(s) = \sqrt{\frac{1}{2\pi}} \int_{-\infty}^{\infty} \cos(xs)\, f(x)\, dx = \sqrt{\frac{2}{\pi}} \int_{0}^{\infty} \cos(xs)\, f(x)\, dx$$

and if f is odd, then the Fourier sine transform becomes

$$(Sf)(s) = \sqrt{\frac{1}{2\pi}} \int_{-\infty}^{\infty} \sin(xs)\, f(x)\, dx = \sqrt{\frac{2}{\pi}} \int_{0}^{\infty} \sin(xs)\, f(x)\, dx.$$

Hereafter we take for granted the Fourier integral theorem for complex Fourier transforms: $F^{-1}Ff = f$ and as its equivalent statements, $L^{-1}Lf = f$ and $M^{-1}Mf = f$ for granted.

We shall give some illustrative examples.

The formula

$$L[\sin(\omega t)](s) = \frac{\omega}{s^2 + \omega^2} \quad (\omega > 0,\ \sigma = \operatorname{Re} s > 0) \tag{7.26}$$

and its inverse

$$L^{-1}\left[\frac{\omega}{s^2 + \omega^2} \right](t) = \sin(\omega t) \tag{7.27}$$

are very important in the application of the theory of Laplace transforms.

The customary proof of (7.26) is the following. By integration by parts, we have

$$\int e^{-st}\sin(\omega t)\,dt = -\frac{1}{s}e^{-st}\sin(\omega t) - \frac{\omega}{s^2}e^{-st}\cos(\omega t) - \frac{\omega^2}{s^2}\int e^{-st}\sin(\omega t)\,dt,$$

whence it follows that

$$\int e^{-st}\sin(\omega t)\,dt = \frac{s^2}{s^2+\omega^2}\left(-\frac{1}{s}e^{-st}\sin(\omega t) - \frac{\omega}{s^2}e^{-st}\cos(\omega t)\right). \quad (7.28)$$

Under the condition $\sigma > 0$, the infinite integral $\int_0^\infty e^{-st}\sin(\omega t)\,dt$ is absolutely convergent, and by (7.28) is equal to

$$\frac{s^2}{s^2+\omega^2}\left[-\frac{1}{s}e^{-st}\sin(\omega t) - \frac{\omega}{s^2}e^{-st}\cos(\omega t)\right]_0^\infty = \frac{\omega}{s^2+\omega^2}.$$

Similarly we may prove the corresponding formula for cosines:

$$L[\cos(\omega t)](s) = \frac{s}{s^2+\omega^2} \quad (\omega > 0,\ \sigma > 0) \quad (7.29)$$

$$L^{-1}\left[\frac{s}{s^2+\omega^2}\right](t) = \cos(\omega t). \quad (7.30)$$

The following proof is, however, much more concise and instructive. Suppose first that $s = \sigma > 0$ and invoke Euler's identity to deduce that

$$L[\sin(\omega t)](s) = \operatorname{Im}\int_0^\infty e^{-st}e^{i\omega t}\,dt$$

$$= \operatorname{Im}\left[-\frac{1}{s-i\omega}e^{-(s-i\omega t)}\right]_0^\infty$$

$$= \operatorname{Im}\frac{1}{s-i\omega t} = \operatorname{Im}\frac{s+i\omega}{s^2+\omega^2} = \frac{\omega}{s^2+\omega^2}.$$

Now, $L[\sin(\omega t)](s)$ is an analytic function in s for $\sigma > 0$ since the integral is absolutely convergent there, and so is the function $\frac{\omega}{s^2+\omega^2}$. Hence, by the principle of analytic continuation (Theorem A.9), they must coincide in the region $\sigma > 0$, and this proves Formula (7.26).

Note that the above argument also gives a proof of (7.29), since

$$L[\cos(\omega t)](s) = \operatorname{Re}\int_0^\infty e^{-st}e^{i\omega t}\,dt = \operatorname{Re}\frac{s+i\omega}{s^2+\omega^2} = \frac{s}{s^2+\omega^2} \quad (\sigma > 0)$$

Other examples are

$$L\left[\operatorname{erfc}\left(at^{-1/2}\right)\right](p) = \frac{1}{p}\,e^{-2a\sqrt{p}} \tag{7.31}$$

where the error function is defined by

$$\Gamma\left(\frac{1}{2}, z\right) = \sqrt{\pi}\,\operatorname{erfc}\left(\sqrt{z}\right), \tag{7.32}$$

or more generally,

$$L\left[\Gamma\left(\nu, \frac{a}{t}\right)\right](p) = 2a^{\frac{\nu}{2}}p^{\frac{\nu}{2}-1}K_\nu(2\sqrt{ap}), \tag{7.33}$$

$K_\nu(z)$ indicating the modified Bessel function of the second kind defined by (6.10), which for $\nu = \frac{1}{2}$ reduces to (7.31) in view of (6.14) and

$$L[\Gamma(\nu, at)](p) = \frac{\Gamma(\nu)}{p}\left(1 - \frac{1}{(1 + \frac{p}{a})^\nu}\right). \tag{7.34}$$

7.3   Fourier transform

In this section, we shall show that the one-sided (complex) Fourier transform and the Laplace transform have the same function by illustrating Examples 7.6 and 7.7. We use the following data and scheme.

One-sided Fourier Transform

Variable $z \in \mathbb{C}$

the region of absolute convergence $\operatorname{Im} z < -\alpha$ ($\alpha \in \mathbb{R}$) if $|y(t)| = O(e^{\alpha t})$.

$$\hat{y}(z) = F_+[y](z)$$
$$= \frac{1}{\sqrt{2\pi}} \int_0^\infty y(t)\, e^{-izt}\, dt.$$

When $y'(t)$ is continuous for $t > 0$ and $\lim_{t \to \infty} y'(t)\, e^{-izt} = 0$

$$F_+[y'](z) = iz F_+[y](z) - \frac{y(+0)}{\sqrt{2\pi}}.$$

Under similar conditions

$$F_+[y''](z) = (iz)^2 F_+[y](z)$$
$$- \frac{1}{\sqrt{2\pi}}(izy(+0) + y'(+0)).$$

$$F_+^{-1}\left[\frac{1}{z-\alpha}\right](t) = \sqrt{2\pi}\, i\, e^{i\alpha t}$$

$$F_+\left[e^{i\alpha t}\right](t) = \frac{1}{\sqrt{2\pi}\, i}\, \frac{1}{z-\alpha}$$

Laplace Transform

Variable $s = iz \in \mathbb{C}$

Region of absolute convergence $\operatorname{Re} s > \alpha$ if $|y| = O(e^{\alpha t})$

$$Y(s) = L[y](s)$$
$$= \int_0^\infty y(t)\, e^{-st}\, dt$$

When $y'(t)$ is continuous for $t > 0$ and $\lim_{t \to \infty} y'(t)\, e^{-st} = 0$,

$$L[y'](s) = sL[y](s) - sy(+0).$$

Under similar conditions

$$L[y''](s) = s^2 L[y](s)$$
$$- sy(+0) - y'(+0).$$

$$L^{-1}\left[\frac{1}{s-\alpha}\right](t) = e^{\alpha t}$$

$$L[e^{\alpha t}](s) = \frac{1}{s-\alpha}$$

Example 7.6 We solve the differential equation

$$y'' + \omega^2 y = a, \tag{7.35}$$

in $y = y(t)$ with initial conditions

$$y(0) = b, \ y'(0) = c,$$

by the Fourier transform method. The above scheme reads:

$$F_+[y''] + \omega^2 F_+[y] = a F_+[1]$$

$$(iz)^2\, \hat{y}(z) - \frac{1}{\sqrt{2\pi}}(izy(+0) + y'(+0)) + \omega^2\, \hat{y}(z) = a\, \frac{1}{\sqrt{2\pi i}}\, \frac{1}{z}$$

$$\left((iz)^2 + \omega^2\right) \hat{y}(z) = \frac{1}{\sqrt{2\pi}\,z}\left(b\,i\,z^2 + cz - ia\right)$$

$$\hat{y}(z) = \frac{1}{\sqrt{2\pi}}\,\frac{-ibz^2 - cz + ia}{z(z-\omega)(z+\omega)} = \frac{1}{\sqrt{2\pi}}\left(\frac{A}{z} + \frac{B}{z-\omega} + \frac{C}{z+\omega}\right)$$

$$A = \lim_{z\to 0} z\hat{y}(z) = \lim_{z\to 0}\frac{-ibz^2 - cz + ia}{(z-\omega)(z+\omega)} = \frac{ia}{-\omega^2} = -i\,\frac{a}{\omega^2}$$

$$B = \lim_{z\to\infty}(z-\omega)\hat{y}(z) = \lim_{z\to\infty}\frac{-ibz^2 - cz + ia}{z(z+\omega)}$$

$$= \frac{-ib\omega^2 - c\omega + ia}{\omega\cdot 2\omega} = -i\,\frac{b\omega^2 - ic\omega - a}{2\omega^2}$$

$$C = \lim_{z\to\infty}(z+\omega)\hat{y}(z) = \lim_{z\to\infty}\frac{-ibz^2 - cz + ia}{z(z-\omega)}$$

$$= \frac{-ib\omega^2 + c\omega + ia}{-\omega\cdot(-2\omega)} = \frac{-ib\omega^2 + c\omega - a}{2\omega^2}$$

$$y = \frac{1}{\sqrt{2\pi}}\left(AF_+^{-1}\left[\frac{1}{z}\right] + BF_+^{-1}\left[\frac{1}{z-\omega}\right] + CF_+^{-1}\left[\frac{1}{z+\omega}\right]\right)$$

$$= \frac{1}{\sqrt{2\pi}}\left(A\sqrt{2\pi}\,i + B\sqrt{2\pi}\,i\,e^{i\omega t} + C\sqrt{2\pi}\,i\,e^{-i\omega t}\right)$$

$$= i\cdot(-i)\left(\frac{a}{\omega^2} + \frac{b\,\omega^2 - ic\omega - a}{2\omega^2}e^{i\omega t} + \frac{b\,\omega^2 + ic\omega - a}{2\omega^2}e^{-i\omega t}\right)$$

$$= \frac{a}{\omega^2} + 2\,\mathrm{Re}\,\frac{b\,\omega^2 - a - ic\omega}{2\omega^2}e^{i\omega t}$$

$$= \frac{a}{\omega^2} + \frac{1}{\omega^2}\left((b\,\omega^2 - a)\cos(\omega t) + c\omega\sin(\omega t)\right),$$

which is the solution.

Example 7.7 We solve the same differential equation (7.35) under the same initial conditions as in Example 7.6 by the Laplace transform method. The above scheme reads.

$$L[y''] + \omega^2 L[y] = aL[1]$$

$$s^2\,Y(s) - s\,y(0) - y'(0) + \omega^2\,Y(s) = a\,\frac{1}{s}$$

$$\left(s^2 + \omega^2\right)Y(s) = b\,s + c + \frac{a}{s}$$

$$Y(s) = \frac{bs^2 + cs + a}{s(s^2 + \omega^2)} = \frac{A}{s} + \frac{B}{s - i\omega} + \frac{C}{s + i\omega}$$

$$A = \lim_{s \to 0} s\, Y(s) = \lim_{s \to 0} \frac{bs^2 + cs + a}{s^2 + \omega^2} = \frac{a}{\omega^2}$$

$$B = \lim_{s \to i\omega} (s - i\omega)\, Y(s) = \lim_{s \to i\omega} \frac{bs^2 + cs + a}{s(s + i\omega)}$$

$$= \frac{b(i\omega)^2 + c(i\omega) + a}{i\omega \cdot 2i\omega} = \frac{b\omega^2 - ic\omega - a}{2\omega^2}$$

$$C = \lim_{s \to -i\omega} (s + i\omega)\, Y(s) = \lim_{s \to -i\omega} \frac{bs^2 + cs + a}{s(s - i\omega)}$$

$$= \frac{-b\omega^2 - ic\omega + a}{-i\omega \cdot (-2i\omega)}$$

$$y = A L^{-1}\left[\frac{1}{s}\right] + B L^{-1}\left[\frac{1}{s - i\omega}\right] + C L^{-1}\left[\frac{1}{s + i\omega}\right]$$

$$= A + B\, e^{i\omega t} + C\, e^{-i\omega t}$$

$$= \frac{a}{\omega^2} + \frac{b\omega^2 - a - ic\omega}{2\omega^2}\, e^{i\omega t} + \frac{b\omega^2 - a + ic\omega}{2\omega^2}\, e^{-i\omega t}$$

$$= \frac{a}{\omega^2} + \frac{1}{\omega^2}\left((b\omega^2 - a)\cos(\omega t) + c\omega \sin(\omega t)\right),$$

which is the solution.

7.4 Mellin transform

In connection with the theory of Fourier and Laplace transforms, we shall state basics of the theory of **Mellin transforms**. If we put

$$e^{-t} = x, \quad f(t) = g(x), \quad F_{II}(s) = G_M(s),$$

in the two-sided Laplace transform

$$(L_{II}f)(s) = \int_{-\infty}^{\infty} e^{-ts} f(t)\, \mathrm{d}t,$$

then we get the **Mellin transform**

$$M(f)(s) = \int_0^{\infty} x^{s-1} f(x)\, \mathrm{d}x$$

and the **inverse Mellin transform**

$$\left(M^{-1}(f)\right)(x) = \frac{1}{2\pi i} \int_{(c)} f(s) x^{-s} \, \mathrm{d}s$$

(where the integral is taken along the vertical Bromwich path $\mathrm{Re}\, s = c$) are called the **pair of its Mellin transforms**. We shall give some typical examples. First, the Mellin transform of $\cos x$ is $\cos\left(\frac{\pi}{2}s\right)\Gamma(s)$, i.e.

$$M(\cos)(s) = \int_0^\infty x^{s-1} \cos x \, \mathrm{d}x = \cos\left(\frac{\pi}{2}s\right)\Gamma(s).$$

To see this it is sufficient to compute $M^{-1}\left(\cos\left(\frac{\pi}{2}s\right)\Gamma(s)\right)(x)$ and we are to show that

$$\cos(x) = \frac{1}{2\pi i} \int_{(c)} x^{-s} \cos\left(\frac{\pi}{2}s\right)\Gamma(s) \, \mathrm{d}s$$

and this can be easily proved by the Residue Theorem (Theorem A.11), the convergence of the integral taken for granted (which can be checked by the Stirling formula (Corollary 5.1))

$$\begin{aligned}
\mathrm{RHS} &= \sum_{m=0}^{\infty} \operatorname*{Res}_{s=-2m} x^{-s} \cos\left(\frac{\pi}{2}s\right)\Gamma(s) = \sum_{m=0}^{\infty} \cos m\pi \frac{(-1)^{2m}}{(2m)!} x^{2m} \\
&= \sum_{m=0}^{\infty} \frac{(-1)^m}{(2m)!} x^{2m} = \cos x,
\end{aligned}$$

where we used (2.4).

Exercise 7.9 Similarly as above prove the following pair of Mellin transforms.

$$M(\sin)(s) = \int_0^\infty x^{s-1} \sin x \, \mathrm{d}x = \sin\left(\frac{\pi}{2}s\right)\Gamma(s),$$

$$\sin(x) = \frac{1}{2\pi i} \int_{(c)} x^{-s} \sin\left(\frac{\pi}{2}s\right)\Gamma(s) \, \mathrm{d}s.$$

$$M\left(\frac{1}{1+\bullet}\right)(s) = \int_0^\infty x^{s-1} \frac{1}{1+x} \, \mathrm{d}x = \frac{\pi}{\sin \pi s},$$

$$\frac{1}{1+x} = \frac{1}{2\pi i} \int_{(c)} x^{-s} \frac{\pi}{\sin \pi s} \, \mathrm{d}s.$$

$$M\left(e^{-\bullet}\right)(s) = \int_0^\infty x^{s-1}e^{-x}\,\mathrm{d}x = \Gamma(s),$$

$$e^{-x} = \frac{1}{2\pi i}\int_{(c)} x^{-s}\,\Gamma(s)\,\mathrm{d}s.$$

The second pair appears in Corollary A.4, while the last pair is the most well-known one appearing as (2.1) above. The improper integrals being all absolutely convergent for large values of Re s for which the above formulas hold. Other examples include (7.23) and (B.5).

Chapter 8

Around Dirichlet's *L*-functions

Abstract

In this chapter we shall state rudiments of harmonic analysis on $\mathbb{Z}/q\mathbb{Z}$ and its applications to number-theoretic problems, i.e. to the problems in relation to associated Dirichlet series with periodic coefficients.

In §8.1 we shall develop the theory of finite Fourier series (or what amounts to the same, finite Fourier transforms) to such an extent that is sufficient for our intended number-theoretic applications (based on [IK], [Ka], [Ya]). It is instructive to state the results in parallel to those in the general theory of Fourier series (transforms) expounded in Chapter 7.

In §8.2, we shall establish a remarkable equivalence between Gauss' formula and a finite expression for the value of the Dirichlet *L*-function at 1 as expounded in [HKT].

8.1 The theory of periodic Dirichlet series

In what follows $q > 1$ always indicates the fixed modulus.

Definition 8.1   The space $C(q)$ of all arithmetic functions $f(n)$ of period q,

$$f : \mathbb{Z} \to \mathbb{C}; \ f(n + q) = f(n), \ n \in \mathbb{Z},$$

Fig. 8.1   Gauss

forms a vector space over $\mathbb{C}$. For $f, g \in C(q)$, we define their inner product

$$(f, g) = \sum_{a \bmod q} f(a)\, \overline{g(a)}.$$

Then $C(q)$ becomes a metric vector space with respect to the norm $\|f\| = (f, f)^{\frac{1}{2}}$.

The notions of orthogonality, ONS (orthonormal system), ONB (orthonormal basis) remain the same except that we now speak of the inner product.

Exercise 8.1   Prove that (1) satisfies the defining properties of a scalar product (cf. Problem 4.4).

Example 8.1   For each residue class j mod q let ε_j be defined by

$$\varepsilon_j(n) = e^{2\pi i \frac{j}{q} n}.$$

Fig. 8.2   Dirichlet

Then $\left\{ \dfrac{1}{q}\varepsilon_0, \ \dfrac{1}{q}\varepsilon_1, \ \cdots, \ \dfrac{1}{q}\varepsilon_{q-1} \right\}$ is an ONB of $C(q)$.

Proof.   This follows from the relation (cf. (8.5))

$$(\varepsilon_j, \ \varepsilon_k) = \sum_{a \bmod q} e^{2\pi i \frac{j-k}{q}} = \begin{cases} q, & j = k \\ 0, & j \ne k. \end{cases}$$

$\square$

Example 8.2   Let χ_j denote the characteristic function of the residue class $j \bmod q$:

$$\chi_j(n) = \begin{cases} 1, & n \equiv j \ (\bmod q) \\ 0, & n \not\equiv j \ (\bmod q). \end{cases}$$

Then $\{\chi_0, \ \chi_1, \ \cdots, \ \chi_{q-1}\}$ is an ONB of $C(q)$.

Definition 8.2   Let $(\mathbb{Z}/q\mathbb{Z})^\times$ denote the multiplicative group of reduced residue classes mod q. Let $\chi \in \left(\widehat{\mathbb{Z}/q\mathbb{Z}} \right)^\times$ be an Abelian character, i.e. a homomorphism into $\mathbb{C}^\times$:

$$\chi(\bar{a}\bar{b}) = \chi(\bar{a})\,\chi(\bar{b})$$

and $\bar{a} = a \bmod q$. We extend the domain of definition of χ by 0-extension, i.e. we define $\chi(a) = \chi(\bar{a})$ for $(a, q) = 1$, and $\chi(a) = 0$ for $(a, q) > 1$. Then this χ is a completely multiplicative periodic function called a Dirichlet character mod q:

$$\chi : \mathbb{Z} \to \mathbb{C}, \ \chi(ab) = \chi(a)\,\chi(b), \ \chi(a + q) = \chi(a).$$

The particular character induced by the identity χ_0 of $\left(\widehat{\mathbb{Z}/q\mathbb{Z}}\right)^\times$ is called the principal character mod q and is denoted by χ_0.

Define the Fourier transform $\hat{f}(j)$ of f with respect to the ONB $\{\varepsilon_j\}$:

$$\hat{f}(j) = \frac{1}{q}\,(f, \varepsilon_j) = \frac{1}{q} \sum_{a \bmod q} f(a)\,\overline{\varepsilon_j(a)}. \tag{8.1}$$

Since $C(q) = \bigoplus_{i=0}^{q-1} \mathbb{C}\varepsilon_j$, we automatically have the Fourier expansion of f:

$$f(n) = \sum_{j=0}^{q-1} \hat{f}(j)\,\varepsilon_j(n) \tag{8.2}$$

(cf. Theorem 7.1) for which (8.1) gives, in particular,

$$\hat{\chi}_j(a) = \frac{1}{q}\,\overline{\varepsilon_a(j)} = \frac{1}{q}\,e^{-2\pi i \frac{a}{q} j} \tag{8.3}$$

and the Fourier expansion of χ_a is

$$\chi_a = \sum_{j \bmod q} \hat{\chi}_j(a)\,\varepsilon_j, \tag{8.4}$$

which is nothing other than the orthogonality of the additive characters ε_j's:

$$\sum_{a=0}^{q-1} \varepsilon_j(a) = \sum_{a=0}^{q-1} e^{2\pi i \frac{j}{q} a} = \begin{cases} q, & j \equiv 0 \ (\bmod\ q) \\ 0, & j \not\equiv 0 \ (\bmod\ q). \end{cases} \tag{8.5}$$

For a Dirichlet character χ, we have (8.32) below. Similarly, the Fourier transform $\tilde{f}$ of f with respect to $\{\chi_j\}$ is

$$\tilde{f}(j) = (f, \chi_j) = \sum_{a \bmod q} f(a) \overline{\chi_j(a)} = f(j). \tag{8.6}$$

E.g.

$$\tilde{\varepsilon}_j(n) = \varepsilon_j(n). \tag{8.7}$$

Hence the Fourier expansion of f is

$$f(n) = \sum_{j=0}^{q-1} \tilde{f}(j) \chi_j(n) = \sum_{j=0}^{q-1} f(j) \chi_j(n), \tag{8.8}$$

which is the same as the decomposition of (positive) integers into residue classes $\bmod\, q$ and holds true for any arithmetic function $f : \mathbb{Z} \to \mathbb{C}$, not necessarily periodic. Especially, in the following form of a series, it is of great importance:

$$\sum_{n=0}^{\infty} f(n) = \sum_{j=0}^{q-1} \sum_{m=0}^{\infty} f(m) \chi_j(m). \tag{8.9}$$

The function $L(s, \chi)$ defined for a Dirichlet character $\chi \bmod q$ by the Dirichlet series

$$L(s, \chi) = \sum_{n=1}^{\infty} \frac{\chi(n)}{n^s}, \quad \sigma > 1 \tag{8.10}$$

absolutely convergent, is called Dirichlet's *L*-function.

For $\chi \neq \chi_0$, (8.10) is (conditionally and uniformly) convergent for $\sigma > 1$ by Corollary B.1, and we may speak of the value $L(1, \chi)$. Formula (8.32) gives for χ primitive (by (8.40))

$$L(s, \chi) = \frac{1}{G(\overline{\chi})} \sum_{k=1}^{q-1} \overline{\chi}(k) \, l_s\left(\frac{k}{q}\right). \tag{8.11}$$

Example 8.3 Consider the general Hurwitz-Lerch zeta-function

$$\Phi(s, a, z) = \sum_{n=0}^{\infty} \frac{z^n}{(n + a)^s}, \tag{8.12}$$

$a > 0$, $\sigma > 1$. Then, by (8.9),

$$\Phi(s, a, z) = \sum_{j=0}^{q-1} \sum_{\substack{m=0 \\ m \equiv j \ (\text{mod } q)}}^{\infty} \frac{z^m}{(n+a)^s}$$

$$= \sum_{j \neq 0}^{q-1} \sum_{m=0}^{\infty} \frac{z^{j+mq}}{(j+mq+a)^s}$$

$$= q^{-s} \sum_{j=0}^{q-1} z^j \sum_{m=0}^{\infty} \frac{(z^q)^m}{\left(m + \frac{j+a}{q}\right)^s},$$

whence

$$\Phi(s, a, z) = q^{-s} \sum_{j=0}^{q-1} z^j \, \Phi\left(s, \frac{j+a}{q}, z^q\right), \tag{8.13}$$

which is the most general Kubert identity (or distribution property).

We introduce the vector space of Dirichlet series isomorphic to $C(q)$:

Definition 8.3 Let $D(q)$ denote the space of (formal) Dirichlet series $\sum_{a=1}^{\infty} \frac{f(n)}{n^s}$, with $f(n) \in C(q)$:

$$D(q) = \left\{ \sum_{n=1}^{\infty} \frac{f(n)}{n^s} \, \middle| \, f \in C(q), \ \sigma \gg 1 \right\}.$$

If we impose some growth condition on f, the Dirichlet series is convergent in some half-plane, $\sigma \gg 1$. The spaces $C(q)$ and $D(q)$ are isomorphic (cf. Yamamoto [Ya]).

We may express (8.8) in terms of the Dirichlet series

$$\sum_{n=1}^{\infty} \frac{f(n)}{n^s} = \sum_{a \bmod q} f(a) \sum_{n=1}^{\infty} \frac{\chi_a(n)}{n^s} \tag{8.14}$$

$$= \sum_{a \bmod q} f(a) \, \zeta(s, a \bmod q),$$

where $\zeta(s, a \bmod q)$ signifies the partial zeta-function

$$\zeta(s, a \bmod q) = \sum_{n \equiv a \bmod q} \frac{1}{n^s}, \tag{8.15}$$

which can be expressed by the Hurwitz zeta-function

$$\zeta(s, a \bmod q) = q^{-s} \zeta\left(s, \frac{a}{q}\right),$$ (8.16)

so that it is meromorphic over the whole plane. Hence (8.12) is stated as

$$\sum_{n=1}^{\infty} \frac{f(n)}{n^s} = q^{-s} \sum_{a \bmod q} f(a) \zeta\left(s, \frac{a}{q}\right),$$ (8.17)

which continues the left-hand side meromorphically over the whole plane.

Example 8.4 If we trace the above argument in the reverse direction, we get

$$q^{-s} \zeta\left(s, \frac{a}{q}\right) = \sum_{n=1}^{\infty} \frac{\chi_a(n)}{n^s}.$$

Substituting (8.4) and (8.3) for χ_a, we obtain

$$q^{-s} \zeta\left(s, \frac{a}{q}\right) = \sum_{n=1}^{\infty} \sum_{j \bmod q} \hat{\chi}_j(a)\, \varepsilon_j(n) \frac{1}{n^s}$$

$$= \frac{1}{q} \sum_{j \bmod q} \overline{\varepsilon_a(j)} \sum_{n=1}^{\infty} \frac{\epsilon_j(n)}{n^s},$$

whence we have

$$q^{-s} \zeta\left(s, \frac{a}{q}\right) = \frac{1}{q} \sum_{j \bmod q} \overline{\varepsilon_a(j)}\, l_s\left(\frac{j}{q}\right),$$ (8.18)

being valid for all $s \neq 1$, whose $l_s(x)$ indicates the polylogarithm function defined by (3.3) (cf. Ishibashi [Is, p.447]).

But, if we state (8.18) in the form (the genuine generalization of the Eisenstein formula)

$$\sum_{k=1}^{q-1} e^{-2\pi i \frac{a}{q} k}\, l_s\left(\frac{k}{q}\right) = q^{1-s} \zeta\left(s, \frac{a}{q}\right) - \zeta(s),$$ (8.19)

and interpret the case $s = 1$ as the limit as $s \to 1$, then (8.19) is valid for all $s \in \mathbb{C}$. The limit interpretation of the right-hand side of (8.19) corresponds to the normalization $l_s(0) = \zeta(s)$ in Milnor [Mi], but even under this, (8.16) is not valid and only (8.19) stands. We state the limiting case of (8.19) as

Theorem 8.1 *The limiting case of (8.19) implies Gauss' formula (8.29).*

For details, cf. [LDH].

Exercise 8.2 Prove that the formula

$$e^{2\pi i j\xi}\, q^{-s}\, \phi\left(s, \frac{a+j}{q}, q\xi\right) = \frac{1}{q}\sum_{r=0}^{q-1} e^{-2\pi i \frac{j}{q} r}\, \phi\left(s, a, \xi + \frac{r}{q}\right) \qquad (8.20)$$

is the two different expressions for the j-component

$$\phi_j(s, a, \xi) = \sum_{m=0}^{\infty} \frac{e^{2\pi i m\xi}}{(m+a)^s}\, \chi_j(m)$$

of

$$\phi(s, a, \xi) = \sum_{j=0}^{q-1} \phi_j(s, a, \xi), \qquad (8.21)$$

where $\phi(s, a, \xi) = \Phi(s, a, e^{2\pi i\xi})$ (cf. (8.12)). Note that one expression is a consequence of (8.4),

$$\chi_a(m) = \frac{1}{q}\sum_{j \bmod q} \overline{\epsilon_a(j)}\, \epsilon_j(m) = \frac{1}{q}\sum_{j \bmod q} e^{2\pi i \frac{m-a}{q} j}. \qquad (8.22)$$

Solution   Since $\phi_j(s, a, \xi) = \displaystyle\sum_{\substack{m=0 \\ m\equiv j\ (\bmod q)}}^{\infty} \frac{e^{2\pi i m\xi}}{(m+a)^s}$, we may express it, on

writing $m = j + lq$, $l = 0, 1, \cdots$, as

$$\phi_j(s, a, \xi) = q^{-s}\sum_{l=0}^{\infty} \frac{e^{2\pi i(lq\xi+j\xi)}}{\left(l+\frac{a+j}{q}\right)^s}$$

$$= e^{2\pi i j\xi} q^{-s}\, \phi\left(s, \frac{a+j}{q}, q\xi\right).$$

On the other hand, substituting (8.3), we derive that

$$\phi_j(s, a, \xi) = \sum_{m=0}^{\infty} \frac{e^{2\pi i m\xi}}{(m+a)^s} \frac{1}{q}\sum_{r=0}^{q-1} e^{2\pi i \frac{m-j}{q} r}$$

$$= \frac{1}{q}\sum_{r=0}^{q-1} e^{-2\pi i \frac{j}{q} r} \sum_{m=0}^{\infty} \frac{e^{2\pi i(\xi+\frac{r}{q})m}}{(m+a)^s},$$

which is the right-hand side of (8.21).

We confirm that adding the right-hand side of (8.21) over j mod q, we obtain $\phi(s, a, \xi)$, i.e. (8.22) in terms of the right-hand side of (8.21).

As suggested by (5.56)′, there is a counterpart of (8.11), which is a decomposition into residue classes mod q:

$$L(s, \chi) = \frac{1}{q^s} \sum_{a=1}^{q-1} \chi(a) \, \zeta\left(s, \frac{a}{q}\right) \tag{8.23}$$

being valid for any Dirichlet character mod q, not necessarily primitive.

Recall the Laurent expansion for $\zeta(s, x)$,

$$\zeta(s, x) = \frac{1}{s-1} - \psi(x) + O(s-1), \quad s \to 1, \tag{8.24}$$

where $\psi(x)$ signifies the Euler digamma function (cf. §5.1)

$$\psi(x) = \frac{\Gamma'}{\Gamma}(x) = (\log \Gamma(x))'. \tag{8.25}$$

Also recall the orthogonality of characters

$$\sum_{a=1}^{q-1} \chi(a) = \begin{cases} 0, & \text{if } \chi \neq \chi_0, \\ \varphi(q), & \text{if } \chi = \chi_0, \end{cases} \tag{8.26}$$

where χ_0 and $\varphi(q)$ stand for the principal character mod q and the **Euler function** defined by

$$\varphi(q) = \sum_{\substack{1 \leq a \leq q \\ (a,q)=1}} 1, \tag{8.27}$$

respectively.

From (8.23), (8.24) and (8.26) we obtain

$$L(s, \chi) = -\frac{1}{q} \sum_{a=1}^{q-1} \chi(a) \, \psi\left(\frac{a}{q}\right) + O(s-1), \quad s \to 1$$

and *a fortiori*

$$L(1, \chi) = -\frac{1}{q} \sum_{a=1}^{q-1} \chi(a) \, \psi\left(\frac{a}{q}\right). \tag{8.28}$$

For the values of $\psi\left(\frac{p}{q}\right)$, we have a formula of Gauss (2.58) which may be stated by Lemma 8.1 below as

$$\psi\left(\frac{p}{q}\right) = -\gamma - \log q - \frac{\pi}{2}\cot\frac{p}{q}\pi + 2\sum_{k\leq\frac{q}{2}}\cos\frac{2pk}{q}\pi\,\log\left(2\sin\frac{k}{q}\pi\right), \quad (8.29)$$

where γ is the Euler constant ($\psi(1) = -\gamma$) ([Böh], [Ca], [GR]).

It was D. H. Lehmer [Leh] who first used (8.29) in his study of generalized Euler constants $\gamma(p,q)$ for an arithmetic progression p mod q. He deduced (8.29) from [Leh, (11)], and the relation [Leh, Theorem 7], between $\gamma(p,q)$ and $\psi\left(\frac{p}{q}\right)$, and stated ([Leh, p.135]) "Our proof via finite Fourier series indicates that Gauss' remarkable result has a completely elementary basis."

Our main purpose is to elaborate on this statement of Lehmer and, on streamlining the argument, to show that (8.29) has a purely number-theoretic basis and ψ is a number-theoretic function. As a converse to this, we shall also put into practice the statement of Deninger [D, p.180], to the effect that (8.29) can be used to evaluate $L(1,\chi)$. Indeed, Funakura was on these lines (cf. [Fu, (1)]) but he appealed to the integral representation of Legendre and applied Lehmer's argument of using $-\log(1 - e^{2\pi i x})$, $0 < x < 1$.

8.2 The Dirichlet class number formula

We may now state our main theorem in this chapter.

Theorem 8.2 *Gauss' formula (8.29) is equivalent to finite expressions for $L(1,\chi)$.*

$$L(1,\chi) = \frac{\pi}{2q}\sum_{a=1}^{q-1}\chi(a)\cot\frac{a}{q}\pi \quad (8.30)$$

for χ odd and

$$L(1,\chi) = -\frac{1}{\sqrt{q}}\sum_{a=1}^{q-1}\widehat{\chi}(a)\log\left(2\sin\frac{a}{q}\pi\right) \quad (8.31)$$

for χ even, where

$$\widehat{\chi}(a) = \frac{1}{q} \sum_{k \bmod q} \chi(k) \, e^{-2\pi i \frac{k}{q} a} \tag{8.32}$$

is the finite Fourier transform of χ (intimately related to the generalized Gauss sum $G(a, \chi)$, see (8.39) below).

Corollary 8.1   *For primitive χ, (8.30) and (8.31) reduce, respectively, to*

$$L(1, \chi_{\text{odd}}) = -\frac{\pi i}{G(\overline{\chi})} \sum_{a=1}^{q-1} \overline{\chi}(a) \, B_1\left(\frac{a}{q}\right) \tag{8.30$'$}$$

and

$$L(1, \chi_{\text{even}}) = -\frac{1}{G(\overline{\chi})} \sum_{a=1}^{q-1} \overline{\chi}(a) \log\left(2 \sin \frac{a}{q} \pi\right), \tag{8.31$'$}$$

where $G(\chi) = G(1, \chi)$ is the normalized Gauss sum.

Remark 8.1   *(i) On symmetry grounds, (8.30) may be stated as*

$$L(1, \chi_{\text{odd}}) = -\frac{\pi i}{\sqrt{q}} \sum_{a=1}^{q-1} \widehat{\chi}(a) \, B_1\left(\frac{a}{q}\right), \tag{8.30$''$}$$

which can be explicitly computed to be (8.30) (cf, e.g. [Fu]).

Although both Funakura [Fu] and Ishibashi-Kanemitsu [IK] treated the case of periodic functions $f(n)$ of period q, the formulas (8.30)$''$ and (8.31) are implicit in Yamamoto's work [Ya], depending on (8.11) and (7.18) & (7.19).

(ii) The last statement of Corollary 8.1 follows, on recalling that the Kronecker characters $\left(\frac{|d|}{\cdot}\right)$ ($d < 0$) and $\left(\frac{d}{\cdot}\right)$ ($d > 0$) are primitive odd and even characters mod $|d|$, respectively.

In the course of proof of Theorem 8.2, we shall encounter an interesting number-theoretic function $\log N_q = -\sum_{d|q} (\mu(d) \log d) \frac{q}{d}$ which eventually cancels out in view of the following Theorem 8.3. We believe this function deserves wider attention and we state

Theorem 8.3 *For $q > 1$, the number-theoretic function $\log N(q) = \log N_q$ admits the following expressions.*

$$\log N_q = -q \sum_{d|q} \frac{\mu(d)}{d} \log d \tag{8.33}$$

$$= -\varphi(q) \sum_{a=1}^{q-1} \log\left(2 \sin\left(\frac{a}{q}\pi \frac{\mu\left(\frac{q}{(a,q)}\right)}{\varphi\left(\frac{q}{(a,q)}\right)}\right)\right) \tag{8.34}$$

$$= \sum_{d|q} \Lambda(d)\, \varphi\left(\frac{q}{d}\right) \tag{8.35}$$

$$= \varphi(q) \sum_{p|q} \frac{\log p}{p-1}, \tag{8.36}$$

the product extending over all prime divisors of q, where μ and Λ signify the Möbius function and the von Mangoldt function, respectively.

For a proof cf. [HKT].

8.3 Proof of the theorems

Let $f(n)$ be an arithmetic periodic function of period q:

$$f : \mathbb{Z} \to \mathbb{C}; \quad f(n+q) = f(n), \quad n \in \mathbb{Z}.$$

We define the parity of f as follows: f is called even if $f(-n) = f(n)$ and odd if $f(-n) = -f(n)$.

We prepare some lemmas, of which Lemma 8.1 is repeatedly used in what follows, without notice.

Lemma 8.1 *If f is odd, then*

$$\sum_{a=1}^{q-1} f(a) = 0$$

and if f is even, then

$$\sum_{a=1}^{q-1} f(a) = 2 \sum_{a \le \frac{q}{2}} f(a) = 2 \sum_{a < \frac{q}{2}} f(a) + \frac{1+(-1)^q}{2} f\left(\frac{q}{2}\right).$$

In particular, if f and χ mod q are of opposite parity, then

$$\sum_{a=1}^{q-1} \chi(a) f(a) = 0$$

while if f and χ are of the same parity and $q > 2$, then

$$\sum_{a=1}^{q-1} \chi(a) f(a) = 2 \sum_{a \leq \frac{q}{2}} \chi(a) f(a) = 2 \sum_{a < \frac{q}{2}} \chi(a) f(a).$$

Lemma 8.2 *The ψ function satisfies Gauss' multiplicative formula or the modified Kubert identity*

$$\psi(x) = \log q + \frac{1}{q} \sum_{a=0}^{q-1} \psi\left(\frac{x+a}{q}\right). \tag{8.37}$$

Lemma 8.3 *Let χ denote a Dirichlet character mod q, $q \geq 3$. Then*

$$\sum_{\chi \text{ even}} \chi(n) = \begin{cases} 0 & \text{if } n \not\equiv \pm 1 \pmod{q}, \\ \dfrac{\varphi(q)}{2} & \text{if } n \equiv \pm 1 \pmod{q}. \end{cases}$$

and

$$\sum_{\chi \text{ odd}} \chi(n) = \begin{cases} 0 & \text{if } n \not\equiv \pm 1 \pmod{q}, \\ \dfrac{\varphi(q)}{2} & \text{if } n \equiv 1 \pmod{q}, \\ -\dfrac{\varphi(q)}{2} & \text{if } n \equiv -1 \pmod{q}, \end{cases}$$

where the sum is extended over all even and odd characters, respectively.

Proof. For $q \geq 3$, the set $\{\pm 1\}$ forms a subgroup of the reduced residue class group $G = (\mathbb{Z}/q\mathbb{Z})^{\times}$ of index 2. Hence the factor group $G/\{\pm 1\}$ has order $\dfrac{\varphi(q)}{2}$. Since the set of all even characters coincides with the character group of $G/\{\pm 1\}$, it follows, from the orthogonality of characters, that

$$\sum_{\chi \text{ even}} \chi(a) = \sum_{\chi \in \widehat{G/\{\pm 1\}}} \chi(n)$$

$$= \begin{cases} 0 & \text{if } n \neq 1 \quad in \quad G/\{\pm 1\}, \\ \dfrac{\varphi(q)}{2} & \text{if } n = 1 \quad in \quad G/\{\pm 1\}, \end{cases}$$

which proves the first assertion. The second assertion follows from the first and the orthogonality relation

$$\sum_{\chi \in \widehat{G}} \chi(n) = \begin{cases} 0 & \text{if } n \not\equiv 1 \pmod{q} \\ \varphi(q) & \text{if } n \equiv 1 \pmod{q}. \end{cases} \tag{8.38}$$

This completes the proof.

It is instructive to give a proof of Corollary 8.1 first. We introduce the generalized Gauss sum

$$G(k, \chi) = \sum_{a=1}^{q-1} \chi(a)\, e^{2\pi i \frac{a}{q} k} \tag{8.39}$$

$$= q\,\chi(-1)\,\widehat{\chi}(k)$$

(cf. (8.32)) and note that it decomposes into

$$G(k, \chi) = \overline{\chi}(k)\, G(\chi) \tag{8.40}$$

if and only if χ is primitive ([Ap4], [Da]). We derive (8.30)′ by appealing to Eisenstein's formula

$$\sum_{k=1}^{q-1} l_0\left(\frac{k}{q}\right) e^{-2\pi i \frac{p}{q} k} = B_1 - q\, B_1\left(\frac{p}{q}\right)$$

or rather its converse (cf. [LDH], [Is], [Wa])

$$\sum_{k=1}^{q-1} B_1\left(\frac{k}{q}\right) e^{2\pi i \frac{p}{q} k} = -l_0\left(\frac{p}{q}\right) - 1 - B_1 \tag{8.41}$$

$$= -\frac{i}{2} \cot \frac{p}{q}\pi.$$

Substituting (8.41) into (8.30), we find that

$$L(1, \chi) = \frac{\pi i}{q} \sum_{k=1}^{q-1} B_1\left(\frac{k}{q}\right) G(k, \chi).$$

Using (8.38) and other known facts

$$G(\overline{\chi}) = \chi(-1)\,\overline{G(\chi)}, \quad |G(\chi)|^2 = q,$$

we conclude (8.30)′.

We may deduce (8.31)′ from (8.31) in a similar way. Substituting (8.32) into (8.31), we obtain

$$L(1,\chi) = -\frac{1}{q} \sum_{k=1}^{q-1} \log\left(2\sin\frac{k}{q}\pi\right) \sum_{a=1}^{q-1} \chi(a) \cos\frac{2\pi k}{q}a \qquad (8.42)$$

whose inner sum is again $G(k,\chi)$. Therefore for χ primitive, we have

$$L(1,\chi) = -\frac{G(\chi)}{q} \sum_{k=1}^{q-1} \overline{\chi}(k) \log\left(2\sin\frac{k}{q}\pi\right),$$

whence (8.31)′ follows in the same way. $\square$

We now turn to the proof of Theorem 8.2.

Proof. That (8.29) implies (8.30) and (8.31) is immediate. Indeed, substituting (8.29) in (8.28) and using Lemma 8.1, we obtain (8.30) for χ odd and (8.42) for χ even, which is the same as (8.31).

Now we are to prove the converse, i.e. we are to deduce (8.29) from (8.30) and (8.31).

With p, $(p,q) = 1$, we multiply (8.28) by $\chi(p^{-1})$ and sum over χ mod q, $\chi \neq \chi_0$ to obtain

$$\sum_{\chi_0 \neq \chi \text{ mod } q} \chi(p^{-1}) L(1,\chi) = -\frac{1}{q} \sum_{a=1}^{q} \psi\left(\frac{a}{q}\right) \sum_{\chi_0 \neq \chi \text{ mod } q} \chi(ap^{-1}) \qquad (8.43)$$

$$= S_1 + S_2,$$

say, where

$$S_1 = -\frac{1}{q} \sum_{a=1}^{q} \psi\left(\frac{a}{q}\right) \sum_{\chi \text{ mod } q} \chi(ap^{-1})$$

and

$$S_2 = \frac{1}{q} \sum_{a=1}^{q} \psi\left(\frac{a}{q}\right) \chi_0(ap^{-1}).$$

By the orthogonality (8.38) of characters,

$$S_1 = -\frac{\varphi(q)}{q} \psi\left(\frac{p}{q}\right). \qquad (8.44)$$

The sum S_2 is

$$S_2 = \frac{1}{q} \sum_{a=1}^{q-1} {}^* \psi\left(\frac{a}{q}\right),$$

the star on the summation sign indicating the sum over all a's, relatively prime to q, $(a, q) = 1$, which condition may be replaced by introducing the sum $\sum_{d|(a,q)} \mu(d)$. Writing the condition $d|(a,q)$ as $d|q$, $a = a'd \leq q-1$, we have

$$S_2 = \frac{1}{q} \sum_{d|q} \mu(d) \sum_{a'=1}^{\frac{q}{d}-1} \psi\left(\frac{a'}{\frac{q}{d}}\right)$$

whose inner sum is $\sum_{a=0}^{\frac{q}{d}} \psi\left(\frac{a+1}{\frac{q}{d}}\right) = -\frac{q}{d} \log \frac{q}{d} - \gamma \frac{q}{d} + \gamma$ by Lemma 8.2. Hence

$$S_2 = -\frac{\log q}{q} \varphi(q) - \frac{1}{q} \log N_q - \frac{\varphi(q)}{q} \gamma, \tag{8.45}$$

where $\log N_q$ is defined by (8.33).

Substituting (8.44) and (8.45) in (8.43), we conclude that

$$\sum_{\chi_0 \neq \chi \bmod q} \chi(p^{-1}) L(1, \chi) = \frac{\varphi(q)}{q} \left(-\psi\left(\frac{p}{q}\right) - \log q - \frac{1}{\varphi(q)} \log N_q - \gamma\right). \tag{8.46}$$

It remains to calculate the left-hand side of (8.46) by dividing the sum into two parts:

$$\sum_{\chi_0 \neq \chi \text{ even}} \quad \text{and} \quad \sum_{\chi \text{ odd}}$$

substituting therewith (8.31) and (8.30), respectively.

First, by (8.31),

$$\sum_{\chi_0 \neq \chi \bmod q} \chi(p^{-1}) L(1, \chi) \tag{8.47}$$

$$= -\frac{1}{\sqrt{q}} \sum_{a=1}^{q-1} \log\left(2 \sin \frac{a}{q}\pi\right) \sum_{\chi_0 \neq \chi \text{ even}} \overline{\chi}(p)\, \widehat{\chi}(a)$$

$$= -\frac{1}{\sqrt{q}} \sum_{a=1}^{q-1} \log\left(2 \sin \frac{a}{q}\pi\right) \sum_{\chi \text{ even}} -\overline{\chi_0}(p)\, \widehat{\chi_0}(a)$$

$$= T_1 + T_2,$$

say, where

$$T_1 = -\frac{1}{\sqrt{q}} \sum_{a=1}^{q-1} \log\left(2 \sin \frac{a}{q}\pi\right) \sum_{\chi \text{ even}} \overline{\chi}(p) \frac{1}{\sqrt{q}} \sum_{k \bmod q} \chi(k)\, e^{-2\pi i \frac{k}{q} a} \tag{8.48}$$

and

$$T_2 = \frac{1}{\sqrt{q}} \sum_{a=1}^{q-1} \log\left(2 \sin \frac{a}{q}\pi\right) \frac{1}{\sqrt{q}} \sum_{k \bmod q} \chi_0(k)\, e^{-2\pi i \frac{k}{q} a}. \tag{8.49}$$

The inner double sum of T_1 is

$$\frac{1}{\sqrt{q}} \sum_{k \bmod q} e^{-2\pi i \frac{k}{q} a} \sum_{\chi \text{ even}} \chi(kp^{-1})$$

$$= \frac{1}{\sqrt{q}} \frac{\varphi(q)}{2} \left(e^{-2\pi i \frac{p}{q} a} + e^{2\pi i \frac{p}{q} a}\right) = \frac{\varphi(q)}{\sqrt{q}} \cos\left(2\pi \frac{p}{q} a\right)$$

by Lemma 8.3, and so

$$T_1 = -\frac{\varphi(q)}{q} \sum_{a=1}^{q-1} \cos\left(2\frac{p}{q} a\pi\right) \log\left(2 \sin \frac{a}{q}\pi\right), \tag{8.50}$$

while the inner sum for T_2 is

$$\sum_{k \bmod q} {}^{*} e^{-2\pi i \frac{k}{q} a},$$

which is equal to

$$\varphi(q) \frac{\mu\left(\frac{q}{(a,q)}\right)}{\varphi\left(\frac{q}{(a,q)}\right)}$$

by Hölder's result (cf. [Leh, p.133]).

Hence

$$T_2 = \frac{\varphi(q)}{q} \sum_{a=1}^{q-1} \log\left(2\sin\frac{a}{q}\pi\right) \frac{\mu\left(\frac{q}{(a,q)}\right)}{\varphi\left(\frac{q}{(a,q)}\right)}$$

but this is $-\frac{1}{q}\log N_q$ by (8.34), i.e.

$$T_2 = -\frac{1}{q}\log N_q. \tag{8.51}$$

Substituting (8.50) and (8.51) in (8.47), we obtain

$$\sum_{\chi_0 \neq \chi \text{ even}} \chi(p^{-1}) L(1,\chi) = -\frac{\varphi(q)}{q} \sum_{a=1}^{q-1} \cos 2\pi\frac{p}{q}a\pi \log\left(2\sin\frac{a}{q}\pi\right) - \frac{1}{q}\log N_q. \tag{8.52}$$

On the other hand, by (8.30) and Lemma 8.3,

$$\sum_{\chi \text{ odd}} \chi(p^{-1}) L(1,\chi) = \frac{\pi}{2q} \sum_{a=1}^{q-1} \cot\frac{a}{q}\pi \sum_{\chi \text{ odd}} \chi(ap^{-1}) \tag{8.53}$$

$$= \frac{\pi}{2q}\left(\frac{\varphi(q)}{2}\cot\frac{p}{q}\pi - \frac{\varphi(q)}{2}\cot\frac{-p}{q}\pi\right)$$

$$= \frac{\pi\varphi(q)}{2q}\cot\frac{p}{q}\pi.$$

Combining (8.52) and (8.53) implies

$$\sum_{\chi_0 \neq \chi \bmod q} \chi(p^{-1}) L(1,\chi) \tag{8.54}$$

$$= -\frac{\varphi(q)}{q} \sum_{a=1}^{q-1} \cos\left(2\frac{p}{q}a\pi\right) \log\left(2\sin\frac{a}{q}\pi\right) - \frac{1}{q}\log N_q + \frac{\varphi(q)}{2q}\pi\cot\frac{p}{q}\pi.$$

Comparing (8.46) and (8.54), we see that the terms involving $\log N_q$ cancel each other and (8.29) follows. This completes the proof. $\qquad\square$

Appendix A

Complex functions

A.1 Function series

Fig. A.1   D'alembert

Theorem A.1   *A uniformly convergent series of analytic functions may be integrated term by term along any curve inside the region of uniform convergence. Namely, if the functions*

$$f_1(z), \ f_2(z), \ ...$$

are analytic in D and the series

$$\sum_{n=1}^{\infty} f_n(z) = f(z)$$

is uniformly convergent in D, then for any curve $C \subset D$, we have

$$\int_C f(z) \, \mathrm{d}z = \int_C \sum_{n=1}^{\infty} f_n(z) \, \mathrm{d}z = \sum_{n=1}^{\infty} \int_C f_n(z) \, \mathrm{d}z.$$

Proof. $f_n(z)$ need not be analytic but enough to be continuous in D (since analyticity $\Rightarrow$ continuity, the assumption is excessive). Since $f(z)$ is continuous in D, it follows that the integral $\int_C f_n(z) \, \mathrm{d}z$, $n \in \mathbb{N}$ exists. So does the integral $\int_C s_n(z) \, \mathrm{d}z$ for $s_n(z) = \sum_{i=1}^{n} f_i(z)$. Since $s_n(z)$ converges to $f(z)$ uniformly on D, we have

$$\forall \varepsilon > 0, \ \exists n_0 = n_0(\varepsilon) \in \mathbb{N} \quad s.t. \quad n > n_0 \ \Rightarrow \ |s_n(z) - f(z)| < \varepsilon, \quad \forall z \in D.$$

Hence for $n \geq n_0$, we have

$$\left| \int_C (s_n(z) - f(z)) \, \mathrm{d}z \right| < \varepsilon \, \Lambda(C),$$

where $\Lambda(C)$ is the length of C so that

$$\lim_{n \to \infty} \int_C s_n(z) \, \mathrm{d}z = \int_C f(z) \, \mathrm{d}z$$

whose left-hand side is nothing other than the definition of

$$\sum_{n=1}^{\infty} \int_C f_n(z) \, \mathrm{d}z.$$

$\square$

Theorem A.2 *The limit of the uniformly convergent series of analytic functions is interchangeable with integration along any curve lying in the region of its uniform convergence. Namely, if*

$$f_1(z), \ f_2(z), \ \ldots$$

are analytic in D and

$$\lim_{n \to \infty} f_n(z) = f(z),$$

uniformly in D, then for any Jordan curve $C \subset D$, we have

$$\int_C f(z)\,\mathrm{d}z = \lim_{n\to\infty} \int_C f_n(z)\,\mathrm{d}z = \int_C \lim_{n\to\infty} f_n(z)\,\mathrm{d}z.$$

Definition A.1 If a sequence (respectively, series) of functions defined on D are uniformly convergent on any bounded closed subset of D (i.e., on any compact subset D' such that $D' \subset D$), we say that the sequence (respectively, series) is **uniformly convergent** on D **in the wide sense.**

Theorem A.3 *If the functions*

$$f_1(z),\ f_2(z),\ \dots$$

are (i) analytic in D and (ii) the series

$$\sum_{n=1}^{\infty} f_n(z)$$

is uniformly convergent in D in the wide sense, then its sum

$$f(z) = \sum_{n=1}^{\infty} f_n(z)$$

is analytic in D and its derivative may be obtained by termwise differentiation:

$$f'(z) = \sum_{n=1}^{\infty} f_n'(z)$$

Also, the termwise differentiated series is uniformly convergent in the wide sense in D.

Corollary A.1 *Any function series*

$$\sum_{n=1}^{\infty} f_n(z) := f(z)$$

that is uniformly convergent in the wide sense in D is termwise differentiable infinitely many times:

$$f^{(k)}(z) = \sum_{n=1}^{\infty} f_n^{(k)}(z), \quad k \in \mathbb{N}.$$

(and the k-times differentiated series is also uniformly convergent in the wide sense in D.)

Proof.   We shall prove both Theorem A.3 and its Corollary A.1 at the same time. By the Cauchy integral formula in Theorem A.7, we have for any rectifiable simple curve C in D and any point z in C,

$$f(z) = \sum_{n=1}^{\infty} f_n(z) = \frac{1}{2\pi i} \sum_{n=1}^{\infty} \int_C \frac{f_n(w)}{w - z}\, dw.$$

But $\sum_{n=1}^{\infty} f_n(w)$ is uniformly convergent on C, and so Theorem A.1 allows us to integrate term by term after multiplying $(w - z)^{-1}$:

$$f(z) = \frac{1}{2\pi i} \int_C \sum_{n=1}^{\infty} \frac{f_n(w)}{w - z}\, dw = \frac{1}{2\pi i} \int_C \frac{f(w)}{w - z}\, dw.$$

Hence the Cauchy integral formula holds for $f(z)$ and so it follows that $f(z)$ is analytic in C and that

$$f^{(k)}(z) = \frac{k!}{2\pi i} \int_C \frac{f(w)}{(w - z)^{k+1}}\, dw, \quad k \in \mathbb{N} \cup \{0\}.$$

Let

$$S_n(z) = \sum_{k=1}^{n} f_k(z)$$

be the n-th partial sum of

$$\sum_{n=1}^{\infty} f_n(z)$$

and take any bounded closed subset D' in D. Then take any simple closed contour $C \subset D$ of finite length containing D' and suppose $dist\,(D', C) = \delta > 0$. The we have

$$S_n^{(k)}(z) = \frac{k!}{2\pi i} \int_C \frac{S_n(w)}{(w - z)^{k+1}}\, dw, \quad k \in \mathbb{N} \cup \{0\}.$$

Hence, it follows that

$$\frac{k!}{2\pi \delta^{k+1}}\, \Lambda(C) \max_{w \in C} |f(w) - S_n(w)|,$$

whence we have

$$\forall \varepsilon > 0, \quad \exists n_0 = n_0\,(\varepsilon) \in N \; s.t. \; n > n_0 \quad \Rightarrow$$

$$|f(w) - S_n(w)| < \varepsilon$$

on C. Hence,

$$\lim_{n \to \infty} S_n^{(k)}(z) = f^{(k)}(z).$$

uniformly on D'.   □

Corollary A.2 (The Weierstrass double series theorem) *Suppose* $\{f_n(z)\}$ *are analytic in* $|z - z_0| < r$ *and has the Taylor expansion*

$$f_n(z) = \sum_{k=0}^{\infty} a_k^{(n)}(z - z_0)^k.$$

Then if

$$\sum_{n=1}^{\infty} f_n(z) = f(z)$$

uniformly in the wide sense on $|z - z_0| < r$, *then* $f(z)$ *is analytic on* $|z - z_0| < r$ *and its Taylor expansion is given by*

$$f(z) = \sum_{k=0}^{\infty} a_k(z - z_0)^k \quad (|z - z_0| < r), \quad a_k = \sum_{n=1}^{\infty} a_k^{(n)}.$$

That is, the iterates of the double series coincide — the order of summation being interchangeable —

$$\sum_{n=1}^{\infty} \sum_{k=0}^{\infty} a_k^{(n)}(z - z_0)^k \left(= \sum_{n=1}^{\infty} f_n(z) = f(z) = \sum_{k=0}^{\infty} a_k(z - z_0)^k \right)$$

Proof.   This is a special case of Theorem A.2. The relation between coefficients follows from the Theorem A.8:

$$a_k = \frac{1}{k!} f^{(k)}(z_0) = \frac{1}{k!} \sum_{n=1}^{\infty} f_n^{(k)}(z_0) = \sum_{n=1}^{\infty} a_k^{(n)}.$$
□

Theorem A.4   *If* $\{f_n(z)\}$ *are analytic in* D *and*

$$\lim_{n \to \infty} f_n(z) = f(z)$$

uniformly in the wide sense on D, *then* $f(z)$ *is again analytic on* D *and*

$$f^{(k)}(z) = \lim_{n \to \infty} f_n^{(k)}(z)$$

uniformly in D *in the wide sense.*

Most of the above results on infinite series apply to infinite integrals in spite of the fact that in the case of partial sums $S_n(z)$, n goes to ∞ discretely while in the case of partial integral $I(b) = \int_0^b f(z)\,dz$, b goes to ∞ continuously.

To assure the uniform convergence in case the series (integrals) are absolutely convergent, the main tool is Weierstrass' M-test (Majorant series test) which in the case of integrals asserts that given a (complex-valued) function $f(x, y)$, $x \in [a, b]$ (resp (a, b) as the case may be), $y \in Y$, if there is a positive (-valued) function $M(x)$ such that for any $y \in Y$,

$$|f(x, y)| \le M(x)$$

and

$$\int_a^\infty M(x)\,dx < \infty \quad (\text{resp.} \int_a^b M(x)\,dx < \infty),$$

then $\int_a^\infty f(x, y)\,dx$ (resp. $\int_a^b M(x)\,dx < \infty$) is absolutely and uniformly convergent on Y.

Following Titchmarsh [Tit], we often refer to this as "by absolute convergence."

If the series or integrals are convergent but not absolutely convergent, i.e. conditionally convergent, we need to appeal to more delicate convergence tests such as Dirichlet's (cf. §B.2).

Exercise A.1   Noting that the principal branch of the natural logarithm $\log z$ (often denoted by $\text{Log}\, z$) may be defined by the Condition

$$\frac{d}{dz} \log z = \frac{1}{z}, \quad \log 1 = 0, \tag{A.1}$$

prove the integral representation ($\text{Re}\, z > 0$)

$$\log z = \int_0^\infty \frac{e^{-t} - e^{-zt}}{t}\,dt. \tag{A.2}$$

Solution   Since the integrand $f(t) = f(tz) = -e^{-t}\dfrac{e^{-(z-1)t} - 1}{t} \to z - 1$ as $t \to 0$, the improper integral $\displaystyle\int_0^\infty f(t)\,dt$ is absolutely convergent. Hence we may differentiate under the integral sign to get

$$\frac{d}{dz} \int_0^\infty f(t)\,dt = \int_0^\infty e^{-zt}\,dt = \frac{1}{z}.$$

Since $f(t, 1) = 0$, we have $\log 1 = \int_0^\infty f(t, 1)\, dt = 0$, and Condition (A.1) is satisfied.

Example A.1 (Power series) Series in the form of an infinite degree polynomial

$$f(z) = \sum_{n=0}^\infty a_n (z - z_0)^n,$$

centered at z_0, is called a power series centered at z_0. By translation it suffices to consider the power series at 0. We speak of absolute convergence of power series, and the region of (absolute) convergence of a power series is a disc with the boundary called the circle of convergence whose radius is called the radius of convergence and is most conveniently given by D'Alembert's formula

$$r = \lim_{n \to \infty} \left| \frac{a_n}{a_{n+1}} \right|$$

if the limit exists (including ∞).

We consider the case where $0 < r \le \infty$ and denote the region of convergence by D.

By above theorems we have

Theorem A.5 *Inside D, the region of convergence, (i) a power series may be integrated term by term along any path lying in D, (ii) a power series is analytic in D and the derivatives may be obtained by term by term differentiation and (iii) two power series $\sum a_n z^n$, $\sum b_n z^n$ (in their common region D of convergence) may be added, subtracted, multiplied and divided; in particular, the multiplication is carried out by the Cauchy product*

$$\left(\sum_{m=0}^\infty a_m z^m \right) \left(\sum_{n=0}^\infty b_n z^n \right) = \sum_{l=0}^\infty c_l z^l,$$

$$c_l = \sum_{m+n=l} a_m b_n$$

(cf. Remark 1.1); the division is carried out exactly as we do with ordinary polynomials: e.g. to check the numerical values of Bernoulli numbers in Example 1.1, we may divide z by $z + \frac{1}{2!} z^2 + \frac{1}{3!} z^3 + \cdots$ to obtain

$$\frac{z}{e^z - 1} = 1 - \frac{1}{2} z + \frac{\frac{1}{6}}{2!} z^2 - \frac{\frac{1}{30}}{4!} z^4 + \cdots$$

Similarly, we may carry out the division in Exercise 5.4.

To sum up, Theorem A.5 says that we may treat power series as ordinary polynomials (, which disposition is due to Euler).

Example A.2 (Dirichlet series)   For an increasing sequence of positive reals λ_n, the series of functions $f_n(s) = \lambda_n^{-s} = e^{-s\log\lambda_n}$, with logarithm indicating the principal branch,

$$\sum_{n=1}^{\infty} \lambda_n^{-s}$$

is called the Dirichlet series. Contrary to the case of power series, the region of absolute convergence of Dirichlet series is a right half-plane and we may speak about the abscissa of absolute convergence, often denote by σ_a.

We have a counterpart of Theorem A.5 for Dirichlet series.

Theorem A.6   *Let σ_a denote the abscissa of absolute convergence of a Dirichlet series $f(s) = \sum_{n=1}^{\infty} \lambda_n^{-s}$. Then in the region $\sigma > \sigma_a$, $f(s)$ is analytic, integrable and differentiable term by term. Two Dirichlet series $f(s) = \sum_{m=1}^{\infty} a_m\, m^{-s}$ and $g(s) = \sum_{n=1}^{\infty} b_n\, n^{-s}$ may be multiplied by the Dirichlet convolution:*

$$f(s)\, g(s) = \sum_{l=1}^{\infty} c_l\, l^{-s}, \quad c_l = \sum_{mn=l} a_m\, b_n.$$

The last sum is often expressed as

$$c_l = \sum_{d\mid l} a_d\, b_{l/d} = \sum_{d\mid l} a_{l/d}\, b_d,$$

with $d\mid l$ meaning that d runs though all positive divisors of l.

Example A.3   (i) The integral in (2.1) defining the gamma function is (absolutely and) uniformly convergent in the wide sense in $\sigma > 0$. (ii) The series (3.2) defining the Riemann zeta-function is (absolutely and) uniformly convergent in the wide sense in $\sigma > 1$.

Proof.   (i) The integral is improper at both end points. We apply Weierstrass' M-test to $e^{-x}x^{\sigma-1}$. Let s lie in the compact region $0 < \sigma_0 \le \sigma \le R$, $|t| \le R$, $R > 0$. Then for $0 < x < 1$, $e^{-x}x^{\sigma-1} < e^{-x}x^{\sigma_0-1} = O(x^{\sigma_0-1})$ and for $x > 1$, $e^{-x}x^{\sigma-1} < e^{-x}x^{R-1} = O(x^{-2})$. Since $\int_0^1 t^{\sigma_0-1}\, dt = O(1)$ and $\int_1^{\infty} t^{-2}\, dt = O(1)$, we conclude the assertion.

(ii) Let $2^{m-1} \leq N < 2^m$. Them

$$\sum_{n=1}^{N} n^{-\sigma} < 1 + 2 \cdot 2^{-\sigma} + 4 \cdot 4^{-\sigma} + \cdots + 2^{m-1}(2^{m-1})^{-\sigma}$$

$$= 1 + 2^{1-\sigma} + (2^{1-\sigma})^2 + \cdots + (2^{1-\sigma})^{m-1}$$

$$< \frac{1}{1 - 2^{1-\sigma}}.$$

Hence $\sum_{n=1}^{\infty} n^{-\sigma} < \infty$, and the series is absolutely convergent for $\sigma > 1$.
We may also apply (B.4) to obtain

$$\sum_{n \leq x} n^{-\sigma} = \frac{1}{\sigma - 1} + \frac{1}{2} + \frac{x^{1-\sigma}}{1 - \sigma} - \sigma \int_{1}^{\infty} \overline{B}_1(w)^{-\sigma-1} \, du + O(x^{-\sigma}), \quad \sigma > 0$$

$$= O(1), \quad \sigma > 1. \qquad \square$$

Theorem A.7 (Goursat) *If $f(z)$ is analytic in a domain D, them it has all orders of derivatives $f^{(k)}(z)$, which are also analytic in D, given by the Cauchy integral formula*

$$\frac{f^{(k)}(z)}{k!} = \frac{1}{2\pi i} \int_C \frac{f(w)}{(w-z)^{k+1}} \, dw$$

$$= \frac{1}{2\pi i} \int_C \frac{d^k}{dz^k} \frac{f(w)}{w-z} \, dw,$$

where C is a closed Jordan curve contained in D.

This theorem of Goursat draws a clear line between analytic functions and real differentiable functions. The requirement that a function is analytic at a point (in the neighborhood of a point) is such a stringent restriction that it already implies the existence of derivatives of all orders.

Theorem A.8 (Cauchy-Taylor) *If $f(z)$ is analytic at z_0, then it can be expanded into the Taylor series in the maximal circle contained in the domain D of analyticity:*

$$f(z) = \sum_{n=0}^{\infty} a_n(z - z_0)^n,$$

where a_n is given by (Theorem A.7)

$$a_n = \frac{f^{(n)}(z_0)}{n!} = \frac{1}{2\pi i} \int_C \frac{d^n}{dz_0^n} \frac{f(w)}{w - z_0} \, dw,$$

C being any closed contour contained in D.

Actual determination of Taylor coefficients may be done by the method of undetermined coefficients.

Theorem A.9 (Consistency Theorem or the Principle of Analytic Continuation) *If two functions $f(z)$ and $g(z)$ are analytic in a domain D and $f(z) = g(z)$ on a subset of D containing an accumulation point, then $f(z) = g(z)$ all over D.*

This theorem often applies when two functions $f(z)$, $g(z)$ coincide on a segment $\subset \mathbb{R}$, in which case we may extend the domain of analytic functions.

Corollary A.3 *If a function of the real variable x is real analytic at x_0, i.e. it has the power series expansion in the neighborhood $|x - x_0| < r$, $r > 0$*

$$f(x) = \sum_{n=0}^{\infty} a_n (x - x_0)^n, \quad a_n = \frac{f^{(n)}(x_0)}{n!},$$

then there is a unique function $f(z)$ analytic on $|z - x_0| < r$ and coinciding with $f(x)$ on $|x - x_0| < r$, which is given by

$$f(z) = \sum_{n=0}^{\infty} \frac{f^{(n)}(x_0)}{n!} (z - x_0)^n.$$

This is called an analytic continuation of $f(x)$. Most of elementary functions have their analytic continuation as examples of Corollary A.3: e^z, $\sin z$, $\cos z$. In a word, given a real power series in x, we get its analytic continuation by changing x by z, the complex variable.

Theorem A.10 (Laurent expansion) *If $f(z)$ is (one-valued and) analytic in the annulus (ring-shaped domain) $D : r < |z - z_0| < R$ $(0 < r < R)$, then we have the Laurent expansion of $f(z)$:*

$$f(z) = \sum_{n=-\infty}^{\infty} a_n (z - z_0)^n, \tag{A.3}$$

where the n-th Laurent coefficient a_n is given by $(r < \rho < R)$

$$
\begin{aligned}
a_n &= \frac{1}{2\pi i} \int_{|z - z_0| = \rho} \frac{f(z)}{(z - z_0)^{n+1}} \, dz \\
&= \frac{1}{2\pi i} \int_{|z - z_0| = \rho} \frac{1}{n!} f(z) \frac{d^n}{dz_0^n} (z - z_0)^{-1} \, dz, \quad n \in \mathbb{Z}
\end{aligned}
$$

The Laurent series (A.3) converges uniformly in any annulus contained in D.

For $n \geq 0$, $a_n = \frac{1}{n!} f^{(n)}(z_0)$ are the Taylor coefficients. The negative power part $\sum_{n=-\infty}^{-1} a_n (z - z_0)^n$ is called the principal part of $f(z)$ at $z = z_0$.

If the principal part is finite, $\sum_{n=-m}^{-1} a_n (z - z_0)^n$, $a_{-m} \neq 0$, say, then $f(z)$ is said to have an m-th pole at $z = z_0$, and the coefficient a_{-1} is called the residue of $f(z)$ at $z = z_0$, denoted by

$$a_{-1} = \text{Res}_{z=z_0} f(z) = \frac{1}{2\pi i} \int_{|z-z_0|=\rho} f(z) \, dz. \tag{A.4}$$

If $z = z_0$ is a pole of order m of $f(z)$, then the residue may be calculated in a similar way as the Taylor coefficients, by the method of undetermined coefficients:

Clearing the denominator, we have

$$(z - z_0)^m f(z) = a_{-m} + a_{-m+1}(z - z_0) + \cdots + a_{-1}(z - z_0)^{m-1} + \cdots .$$

Hence differentiating $m - 1$ times, we get

$$\frac{d^{m-1}}{dz^{m-1}}((z - z_0)^m f(z)) = (m - 1)! \, a_{-1} + O(z - z_0).$$

Hence

$$\text{Res}_{z=z_0} f(z) = a_{-1} = \frac{1}{(m - 1)!} \lim_{z \to z_0} \frac{d^{m-1}}{dz^{m-1}}((z - z_0)^m f(z)) \tag{A.5}$$

which is applicable to many similar settings. It is advisable to remember the process rather than Formula (A.5).

A.2 Residue theorem and its applications

Theorem A.11 (The (Cauchy) Residue Theorem) Let C be a piecewise smooth (Jordan) curve. Suppose $f(z)$ is analytic in a region D containing C except for finitely many singularities (which we may suppose are poles) $z_1, z_2, \cdots, z_n$ $(n = 0$ inclusive). Then we have

$$\int_C f(z) \, dz = 2\pi i \sum_{i=1}^{n} \text{Res}_{z=z_i} f(z). \tag{A.6}$$

Remark A.1 By Theorem A.11, the value of the integral may be determined by computing the residues, which, as stated above, amounts to *clearing the denominator. The vacuous case $n = 0$ implies the most fundamental Cauchy Integral Theorem which asserts that the integral along a closed curve contained in the region of analyticity is 0, which in turn originates from the fact that in this case the region can be made to shrink into a point, a topological feature of analytic functions in a region.*

Theorem A.12 *Suppose $f(z)$ is a rational function in z satisfying the conditions (i) it has no poles on the positive real axis and (ii) for some $a \in R$, $a \notin Z$, we have*

$$\lim_{z \to 0} z^{a+1} f(z) = 0, \quad \lim_{z \to \infty} z^{a+1} f(z) = 0.$$

Then,

$$\int_0^\infty x^a f(x) \, dx = \frac{2\pi i}{1 - e^{2\pi i a}} \sum_{z \neq 0} \text{Res} \left(z^a \, f(z) \right), \tag{A.7}$$

where the power function is defined by $z^a = \exp(a \operatorname{Log} z)$, $\operatorname{Log} z$ signifying the principal branch.

Proof.   First note that the improper integral (A.7) is absolutely convergent both at 0 and ∞ by the Weierstrass M-test. Since $f(z)$ has only finitely many poles, we may choose $0 < r < R$ such that all the poles of $f(z)$ other than the origin lies in the annulus $r < |z| < R$. Let D denote this annulus with branch cut along the positive real axis, i.e. its boundary consisting of the curves C_1: starting from r and moving along the upper edge of the positive real axis to R, moving along the bigger circle C_R and returning back to the point R, then moving along the lower edge of the positive real axis to r, moving along the smaller circle c_r, and returning to the starting point r. C_1 : $z = x$, $0 \leq x \leq R$; C_R : $z = Re^{i\theta}$, $0 < \theta < 2\pi$; C_2 : $z = xe^{2\pi i}$, $x : R \to r$; c_r : $z = re^{-i\theta}$, $0 < \theta < 2\pi$. Then we apply the residue theorem to this cut region. Since the argument of z increases from 0 to 2π, we have

$$\int_{C_2} z^a \, f(z) \, dz = \int_R^r x^a \, e^{2\pi i a} \, f(x) \, dx,$$

whence it follows that

$$\left(1 - e^{2\pi i a}\right) \int_0^\infty x^a \, f(x) \, dx = 2\pi i \sum_{z \neq 0} \text{Res} \left(z^a \, f(z) \right).$$

We shall assign a precise meaning to this procedure. We integrate the branch

$$g(z) = z^a f(z), \quad 0 < \arg z < 2\pi$$

(with branch cut along the positive real axis as above) along C_R and c_r:

$$\int_{C_R} g(z)\,dz, \quad \int_{c_r} g(z)\,dz.$$

By dividing the annulus by any ray starting from the origin and lying inside the second and the third quadrants, we introduce two regions D_1 and D_2 with branch cut along the negative and positive imaginary axis, respectively. For concreteness' sake, we choose the negative real axis (any ray can do if there are no poles of the integrand on it):

$$L : \ z = xe^{\pi i}, \ x : R \to r$$

and $-L$. Now integrate the branch

$$g_1(z) = z^a f(z), \quad z \neq 0, \ -\frac{1}{2}\pi < z < \frac{3}{2}\pi$$

along $\partial D_1 = C_1 + C_{R,1} + L + c_{r,1}$, to get

$$\int_r^R e^{a(\operatorname{Log} x + i0)} f(x)\,dx + \int_{C_{R,1}} g_1 + \int_L g_1 + \int_{c_{r,1}} g_1$$
$$= 2\pi i \operatorname*{Res}_{z \in D_1} e^{a(\log|z| + i \arg z)} f(z),$$

where $C_{R,1}$ (resp. $c_{r,1}$) signifies the upper half of C_R, (resp. c_r,) . Also integrating the branch

$$g_2(z) = z^a f(z), \quad z \neq 0, \ \frac{1}{2}\pi < z < \frac{5}{2}\pi$$

along

$$\partial D_2 = C_2 + C_{R,2} + (-L) + c_{r,2},$$

to get

$$-\int_r^R e^{a(\operatorname{Log} x + i0)} f(x)\,dx + \int_{C_{R,2}} g_2 + \int_{-L} g_2 + \int_{c_{r,2}} g_2$$
$$= 2\pi i \operatorname*{Res}_{z \in D_2} e^{a(\log|z| + i \arg z)} f(z),$$

where $C_{R,2}$ (resp. $c_{r,2}$) signifies the lower half of C_R, (resp. c_r,). Now note that except on the positive real axis,

$$g_1(z) = g(z), \quad D_1 \cup \partial D_1$$

and

$$g_2(z) = g(z), \quad D_2 \cup \partial D_2,$$

so that

$$\int_C + \int_{c_r} + \left(1 - e^{2\pi i a}\right) \int_r^R e^{a \log x} f(x) \, dx = \sum_{z \in D} \operatorname{Res} z^a \, f(a),$$

where D is the union of D_1 and D_2 with L and $-L$ removed:

$$D = (D_1 \cup D_2) - L \cup (-L).$$

Now

$$\int_{C_{R4}} = \int_0^{2\pi} R^a e^{ia\theta} f\left(Re^{i\theta}\right) i Re^{i\theta} \, d\theta = O\left(\int_0^{2\pi} R^{a+1} \left|f\left(Re^{i\theta}\right)\right| d\theta\right)$$

and

$$\int_{c_r} = \int_{2\pi}^0 r^a e^{ia\theta} f\left(re^{i\theta}\right) i re^{i\theta} \, d\theta = O\left(\int_0^{2\pi} R^{a+1} \left|f\left(Re^{i\theta}\right)\right| d\theta\right).$$

whence as $r \to 0+$ and $R \to \infty$, we obtain

$$\int_{\partial D} z^a f(z) \, dz \to \left(1 - e^{2\pi i a}\right) \int_0^\infty x^a f(x) \, dx$$

thereby completing the proof.   □

Corollary A.4 *For $0 < \operatorname{Re} a < 1$ we have*

$$\int_0^\infty \frac{x^{-a}}{x+1} \, dx = \frac{\pi}{\sin \pi a}.$$

We note that Exercise 2.3 could be thought of as giving the value if we assume those formulas appearing there.

Appendix B

Summation formulas and convergence theorems

B.1 Summation formula and its applications

Lemma B.1 (The General Newton-Leibniz Principle) *Suppose* $f :$ $[a, b] \to \mathbb{R}$ *is differentiable on* $[a, b]$ *except for a finite number of points and that* $f'(x) = 0$ *(at endpoints we assume* $\frac{d^-}{dx} f(a), \frac{d^+}{dx} f(a)$ *exist.) Further, suppose* $f(x)$ *is continuous on* $[a, b]$. *Then* f *is a constant on* $[a, b]$.

Proof. If $f(x)$ is differentiable at each point of $(c_0, c_2) \subset [a, b]$ except for c_1, then by the standard Newton-Leibniz Principle, we have $f(x) = C_1$ $(c_0 < x < c_1)$, and $f(x) = C_2$ $(c_1 < x < c_2)$. By continuity, we obtain $C_1 = \lim_{x \to c_1 - 0} f(x) = \lim_{x \to c_1 + 0} f(x) = C_2$. Therefore, f is a constant on (c_0, c_2). In case f is not differentiable at the endpoint a, by the continuity at a, we have $f(a) = \lim_{x \to a + 0} f(x) = C_1$. $\qquad\square$

Theorem B.1 (Abel Summation Formula) *If* $f(t) \in C^1[a, x]$, *then putting* $A(t) = \sum_{a < n \leq t} a(n)$,

$$\sum_{a < n \leq x} a(n) f(n) = A(x) f(x) - \int_a^x A(t) f'(t) \, dt. \qquad (B.1)$$

Proof. (The first proof due to Arhipov and Chubarikov.) For $x > a(\geq 0)$, put

$$F(x) = \sum_{a < n \leq x} a(n) f(n) - A(x) f(x),$$

$$G(x) = -\int_a^x A(t) f'(t) \, dt.$$

Then for $x = m \in \mathbb{Z}$ we have

$$\lim_{x \to m+0} F(x) = \sum_{a < n \leq m} f(n)\, a(n) - f(m)\, A(m).$$

On the other hand,

$$\begin{aligned}
\lim_{x \to m-0} F(x) &= \sum_{a < n \leq m-1} f(n)\, a(n) - f(m)\, A(m-1) \\
&= \sum_{a < n \leq m} f(n)\, a(n) - f(m)\, a(m) - f(m)\, A(m-1) \\
&= \sum_{a < n \leq m} f(n)\, a(n) - f(m)\,(a(m) + A(m-1)) \\
&= \sum_{a < n \leq m} f(n)\, a(n) - f(m)\, A(m)
\end{aligned}$$

whence it follows that $F(x)$ and $G(x)$ are continuous on $[a, x]$ and piecewise differentiable on $[a, x]$. Furthermore

$$F'(x) = G'(x), \quad x \notin \mathbb{Z}.$$

Hence, Lemma B.1 applies, and we have $F(x) = G(x) + C$. And since $F(a) = G(a) = 0$, it follows that $F(x) = G(x)$, i.e. (B.1) ensues. $\qquad \square$

The second proof uses a special case of the formula for integration by parts in the theory of Stieltjes integrals stated in the following theorem:

Theorem B.2 *Suppose $f(t)$ is continuous on $[a, b]$ and that $\alpha(t)$ is of bounded variation. Then the Stieltjes integral $\int_a^b f(t)\, d\alpha(t)$ exists. Further, if $f(t)$ is of bounded variation and $\alpha(t)$ is continuous, then $\int_a^b \alpha(t)\, df(t)$ exists and the formula for integration by parts*

$$\int_a^b f(t)\, d\alpha(t) = \Big[f(t)\alpha(t)\Big]_a^b - \int_a^b \alpha(t)\, df(t) \tag{B.2}$$

holds true.

Proof. (The second proof.) Putting $\alpha(t) = A(t) = \sum_{a < n \leq t} a(n)$, $\alpha(t)$ is a step function and so of bounded variation. Since $\sum_{a < n \leq x} a(n)\, f(n) = \int_a^x f(t)\, d\alpha(t)$, it follows from (B.2) that

$$\sum_{a < n \leq x} a(n)\, f(n) = \int_a^x f(t)\, dA(t) = \Big[f(t)\, A(t)\Big]_a^x - \int_a^x A(t)\, df(t). \tag{B.3}$$

Since $f \in C^1$, the last integral may be written as $\int_a^x A(t) f'(t)\, dt$, and (B.3) leads to (B.1). $\qquad\qquad\qquad\qquad\qquad\qquad\qquad\qquad\qquad\square$

The Stieltjes integral $\int_a^b f(t)\, d\alpha(t)$ exists under weaker conditions than those stated in Theorem B.2.

Theorem B.3 *If $f(t)$ and $\alpha(t)$ are both of bounded variation on $[a, b]$ and have no common discontinuity, then each one is integrable with respect to the other from a to b.*

Proof. (The third proof of Theorem B.1.) Substituting $a(n) = A(n) - A(n-1)$, we see that

$$\text{LHS} = \sum_{a < n \leq x} A(n)\, f(n) - \sum_{a-1 < m \leq x-1} A(m)\, f(m+1)$$

$$= -A([a])\, f([a]) + \sum_{n=[a]+1}^{[x]-1} A(n)\, (f(n) - f(n+1)) + A(x)\, f([x]),$$

where as usual $[x]$ is the integral part of x. Using $f(n) - f(n+1) = -\int_n^{n+1} f'(t)\, dt$ and noting that $A(t) = A(n)$, $n < t < n+1$, we obtain

$$\text{LHS} = -\sum_{n=[a]+1}^{[x]-1} \int_n^{n+1} A(t)\, f'(t)\, dt + A(x)\, f([x])$$

$$= -\int_{[a]+1}^{[x]} A(t)\, f'(t)\, dt + A(x)\, f([x]).$$

We rewrite the RHS as

$$-\int_a^x A(t)\, f'(t)\, dt + \int_a^{[a]+1} A(t)\, f'(t)\, dt + \int_{[x]}^x A(t)\, f'(t)\, dt + A(x)\, f([x])$$

$$= A(x)\, f(x) - \int_a^x A(t)\, f'(t)\, dt.$$

$$\qquad\qquad\qquad\qquad\qquad\qquad\qquad\qquad\qquad\qquad\qquad\qquad\square$

Proof. (The fourth proof.) We transform the integral on the RHS of (B.1) by changing the order of summation and integration,

$$-\int_a^x A(t)\, f'(t)\, \mathrm{d}t = -\int_a^x \sum_{a<n\le t} a(n)\, f'(t)\, \mathrm{d}t$$

$$= -\sum_{a<n\le x} a(n) \int_n^x f'(t)\, \mathrm{d}t$$

$$= -\sum_{a<n\le x} a(n)\big(f(x) - f(n)\big)$$

$$= -f(x) \sum_{a<n\le x} a(n) + \sum_{a<n\le x} a(n)\, f(n),$$

which leads to (B.1). □

Theorem B.4 (Euler's summation formula) *Let* $\overline{B_1}(x) = B_1\,(x - [x]) = x - [x] - \frac{1}{2}$ *denote the 1-st periodic Bernoulli polynomial (cf. (7.9)). Then for $f(t) \in C^1\,([a,x])$, we have*

$$\sum_{a<n\le x} f(n) = \int_a^x f(t)\, \mathrm{d}t - \Big[\overline{B_1}(t)\, f(t)\Big]_a^x + \int_a^x \overline{B_1}(t)\, f'(t)\, \mathrm{d}t.$$

Proof. Putting $a(n) = 1$ in (B.1), we have

$$A(t) = \sum_{a<n\le t} 1 = [t] - [a] = t - \overline{B_1}(t) - \big(a - \overline{B_1}(a)\big).$$

Hence (B.1) reads

$$\sum_{a<n\le x} f(n)$$

$$= \big\{x - \overline{B_1}(x) - (a - \overline{B_1}(a))\big\}\, f(x) - \int_a^x t\, f'(t)\, \mathrm{d}t$$

$$+ \big(a - \overline{B_1}(a)\big)\, (f(x) - f(a)) + \int_a^x \overline{B_1}(t)\, f'(t)\, \mathrm{d}t$$

$$= \Big[(t - \overline{B_1}(t))\, f(t)\Big]_a^x - \Big[t\, f(t)\Big]_a^x + \int_a^x f(t)\, \mathrm{d}t + \int_a^x \overline{B_1}(t)\, f'(t)\, \mathrm{d}t,$$

which leads to the RHS. □

Proof. (a là Arhipov and Chubarikov) Putting

$$F(x) = \sum_{a<n\le x} f(n) - \overline{B_1}(a)\, f(a) + \overline{B_1}(x)\, f(x)$$

and

$$G(x) = \int_a^x \left(f(t) + \overline{B_1}(t) \, f'(t) \right) dt,$$

we shall show that $F(x) = G(x)$. Clearly $F(a) = G(a) = 0$. $G(x)$, being a function of the upper limit of integration. is continuous and so is $F(x)$ for $x \notin \mathbb{Z}$ by definition. And for $x = m \in \mathbb{Z}$, we note that

$$\lim_{x \to m+0} F(x) = \sum_{a < n \le m} f(n) - \overline{B_1}(a) f(a) - \frac{1}{2} f(m) \, (= F(m)) = \lim_{x \to m-0} F(x),$$

i.e. $F(x)$ is continuous for $x = m \in \mathbb{Z}$. This is because the positive jump of $F_1(x) = \sum_{a < n \le x} f(n)$ at $x = m$ cancels the negative jump of $F_2(x) = \overline{B_1}(a)f(a) - \overline{B_1}(x)f(x)$, and their sum $F(x)$ is continuous at $x = m$. Also it is clear that at $x \notin \mathbb{Z}$, both $G(x)$ and $F(x)$ are differentiable and $G'(x) = F'(x)$. Hence the Newton-Leibniz principle in Lemma B.1 applies. □

Theorem B.5 (the Euler-Maclaurin summation formula [Wal]. A general form of Theorem B.4) *Let $\overline{B}_r(x) = B_r(x - [x])$ denote the r-th periodic Bernoulli polynomial. Then for $f(t) \in C^l[a, x]$, we have*

$$\sum_{a < n \le x} f(n) = \int_a^x f(t) \, dt + \sum_{r=1}^l \frac{(-1)^r}{r!} \left\{ \overline{B_r}(x) \, f^{(r-1)}(x) - \overline{B_r}(a) \, f^{(r-1)}(a) \right\}$$

$$+ \frac{(-1)^{l+1}}{l!} \int_a^x \overline{B_l}(t) \, f^{(l)}(t) \, dt.$$

Proof. (The first proof.) Apply integration by parts to Theorem B.4. □

Proof. (The second proof.) (Similar to the second proof of Theorem B.4) Putting

$$F(x) = \sum_{a < n \le x} f(n) - \sum_{r=1}^l \frac{(-1)^r}{r!} \left\{ \overline{B_r}(x) \, f^{(r-1)}(x) - \overline{B_r}(a) \, f^{(r-1)}(a) \right\}$$

and

$$G(x) = \int_a^x \left(f(t) + \frac{(-1)^l}{l!} \overline{B_l}(t) \, f^{(l)}(t) \right) dt,$$

we note that $F(a) = G(a) = 0$, and $F'(x) = G'(x)$, $\forall x \notin \mathbb{Z}$. □

B.2   Application to the Riemann zeta-function

We illustrate our theory in §B.1 by specifying to the Riemann zeta-function.

Theorem B.4 with $a = 1$, $x > 0$, $f(u) = u^{-s}$, $s \in \mathbb{C}$ gives, on noting $\overline{B}_1(1) = -\frac{1}{2}$, $B_1(1) = \frac{1}{2}$,

$$\sum_{1 < n \leq x} n^{-s} = \int_1^x u^{-s} \, du - \left[\overline{B}_1(u) \, u^{-s} \right]_1^x + \int_1^x \overline{B}_1(u) \left(-s \, u^{-s-1} \right) du$$

$$= \frac{x^{1-s}}{1-s} + \frac{1}{s-1} - \overline{B}_1(x) \, x^{-s} - \frac{1}{2} - s \int_1^x \overline{B}_1(u) \, u^{-s-1} \, du,$$

which may be expressed for $\sigma > 0$ as

$$\sum_{n \leq x} n^{-s} = \frac{1}{s-1} + \frac{1}{2} + \frac{x^{1-s}}{1-s} - \overline{B}_1(x) \, x^{-s} - s \int_1^x \overline{B}_1(u) \, u^{-s-1} \, du$$

$$= \frac{1}{s-1} + \frac{1}{2} + \frac{x^{1-s}}{1-s} - s \int_1^\infty \overline{B}_1(u) \, u^{-s-1} \, du + O(x^{-\sigma}). \quad \text{(B.4)}$$

For $\sigma > 1$, letting $x \to \infty$, we obtain

$$\zeta(s) = \frac{1}{s-1} + \frac{1}{2} - s \int_1^\infty \overline{B}_1(u) \, u^{-s-1} \, du. \quad \text{(B.5)}$$

We note that the RHS is meromorphic in $\sigma > 0$, providing a meromorphic continuation of the LHS to $\sigma > 0$. We now show that the integral is analytic for $\sigma > -1$. For this it suffices to show that $\int_1^\infty \overline{B}_1(u) \, u^{-s-1} \, du$ is uniformly convergent. To this end we may apply the mean value theorem or integration by parts:

$$\int_1^\infty \overline{B}_1(u) \, u^{-s-1} \, du = \left[\frac{1}{2} \overline{B}_2(u) \, u^{-s-1} \right]_1^\infty + \frac{s+1}{2} \int_1^\infty \overline{B}_2(u) \, u^{-s-2} \, du.$$

The first term is $-\frac{1}{2} B_2$ and the second term is absolutely convergent for $\sigma > -1$, whence $\int_1^\infty \overline{B}_1(u) \, u^{-s-1} \, du$ is analytic for $\sigma > -1$ and it follows that (B.5) holds and $\zeta(s)$ is analytic for $\sigma > -1$ except for $s = 1$, where it has a simple pole with residue 1.

Now if we restrict to $-1 < \sigma < 0$, then

$$s \int_0^1 \overline{B}_1(u) \, u^{-s-1} \, du = s \int_0^1 \left(u - \frac{1}{2} \right) u^{-s-1} \, du = -\frac{1}{s-1} - \frac{1}{2}.$$

Hence (B.5) is transformed into

$$\zeta(s) = -s \int_0^\infty \overline{B}_1(u) \, u^{-s-1} \, du, \quad -1 < \sigma < 0, \tag{B.6}$$

where the RHS of (B.6) is meromorphic in $-1 < \sigma$. We substitute the boundedly convergent Fourier series (7.9) for $\overline{B}_1(x)$.

Lebesgue's theorem allows us to integrate term by term to obtain:

$$\zeta(s) = \frac{s}{\pi} \int_0^\infty \sum_{n=1}^\infty \frac{\sin 2\pi n u}{n} u^{-s-1} \, du$$

$$= \frac{s}{\pi} \sum_{n=1}^\infty \frac{1}{n} \int_0^\infty u^{-s-1} \sin 2\pi n u \, du. \tag{B.7}$$

We apply a formula in the theory of Mellin transforms (cf. §7.4)

$$\int_0^\infty x^{z-1} \sin ax \, dx = a^{-z} \Gamma(z) \sin\left(\frac{1}{2}\pi z\right), \quad -1 < \mathrm{Re}\, z < 1. \tag{B.8}$$

Substituting (B.8) in (B.7) we deduce that

$$\zeta(s) = \frac{1}{\pi} (2\pi)^s \, s \, \Gamma(-s) \sin\left(-\frac{\pi}{2} s\right) \sum_{n=1}^\infty \frac{1}{n} n^s$$

$$= 2 (2\pi)^{s-1} \Gamma(1-s) \sin\left(\frac{\pi}{2} s\right) \zeta(1-s) \tag{B.9}$$

for $-1 < \sigma < 0$, where we used (2.5). For $\sigma < 0$, $\zeta(1-s)$ has the Dirichlet series expression. Formula (B.9) is the asymmetric form of the functional equation (5.54) for the Riemann zeta-function.

Example B.1 Find the value of $\zeta(2)$.

Solution Applying (2.15) to rewrite (B.9) as

$$\zeta(s) = (2\pi)^{s-1} \frac{\pi}{\Gamma(s) \cos \frac{\pi}{2} s} \zeta(1-s) \tag{B.10}$$

Putting $s = 2$ in (B.10), we get

$$\zeta(2) = 2\pi \frac{\pi}{\Gamma(2)\,(-1)} \zeta(-1) = 2\pi^2 \left(-\zeta(-1)\right)$$

and we are led to find the value of $\zeta(-1)$, which we may find with the aid of (4.3).

Theorem B.6 (Abel's continuity theorem)   *Suppose the power series*
$$f(z) = \sum_{n=0}^{\infty} a_n z^n \text{ converges at the point } z_0 \text{ on the circle of convergence.}$$
Then as long as z approaches z_0 in the angular region $|\arg(z - z_0)| < \dfrac{\pi}{2}$,
the value $f(z)$ approaches the value $\displaystyle\sum_{n=0}^{\infty} a_n z_0^n$:

$$\lim_{z \to z_0} f(z) = \sum_{n=0}^{\infty} a_n z_0^n.$$

Theorem B.7 (Generalization of Dirichlet's theorem)   *Given two*
sequences of functions $\{a_n(x)\}$, $\{b_n(s)\}$ defined on $R \subset \mathbb{R}$ and $D \subset \mathbb{C}$,
respectively, suppose that

(i) the partial sums $A_N(x) = \displaystyle\sum_{n \leq N} a_n(x)$ of the series $\displaystyle\sum_{n=1}^{\infty} a_n(x)$ are
bounded uniformly in x
(ii) $b_n(s) \to 0$ uniformly on D
and that
(iii) there is a Majorant series $\displaystyle\sum_{n=1}^{\infty} c_n < \infty$ of positive terms c_n such
that

$$|b_n(s) - b_{n+1}(s)| \leq c_n$$

for every $s \in D$ and for all n sufficiently large.
Then the series $\sum_{n=1}^{\infty} a_n(x) b_n(s)$ is uniformly convergent on R and D.
If $b_n(s)$ are analytic in D, so is the sum function $\sum_{n=1}^{\infty} a_n(x) b_n(s)$ for
each $x \in R$.

Proof.   By the formula for partial summation (cf. the third proof of Theorem B.3.), we have for integers M, N, $M < N$,

$$\sum_{M < n \leq N} a_n(x) b_n(s) = \sum_{M < n \leq N} A_n(x) (b_n(s) - b_{n+1}(s))$$
$$+ A_N(x) b_N(s) - A_M(x) b_{M+1}(s)$$
$$= O\left(\sum_{M < n \leq N} c_n \right) + O(|b_N(s)| + |b_{M+1}(s)|)$$

which can be made $< \varepsilon$ for any $\varepsilon > 0$ uniformly in x and s provided that M and N are sufficiently large. Hence the Cauchy criterion applies and uniform convergence follows. □

For fixed $x \in R$, analyticity of the sum function is a consequence of Theorem A.3

Corollary B.1 (Dirichlet's test for uniform convergence)   *If $b_n \to 0$ as $n \to \infty$, and $\left| \sum\limits_{n=1}^{N} a_n(x) \right| = O(1)$ uniformly in $x \in R$, then the series $\sum\limits_{n=1}^{\infty} b_n \, a_n(x)$ is uniformly convergent on R.*

Example B.2   If $b_n \to 0$, then $\sum\limits_{n=1}^{\infty} b_n \sin 2\pi n x$ is uniformly convergent in any interval not containing an integer. This follows from Exercise 7.8, (7.14). In particular, the case $b_n = \frac{1}{n}$ establishes the uniform convergence of the Fourier series $-\dfrac{1}{\pi} \sum\limits_{n=1}^{\infty} \dfrac{\sin 2\pi n x}{n}$ for $\overline{B_1}(x)$ in any interval not containing an integer. Therefore, Lebesgue's theorem allows to integrate it term by term to obtain $\overline{B_2}(x)$ (and higher order periodic Bernoulli polynomials).

This example is a special case of the following.

Proposition B.1   *The series for the polylogarithm function $l_s(x)$ defined in the first instance for $\sigma > 1$ by*

$$l_s(x) = \sum_{n=1}^{\infty} \frac{e^{2\pi i n x}}{n^s}$$

is uniformly convergent in any interval of x free from an integer $(0 < x < 1)$ for $\sigma > 0$.

Proof.   By Exercise 7.8, the partial sums satisfy

$$\left| \sum_{k=1}^{n} e^{2\pi i k x} \right| = \left| e^{\pi i (n+1) x} \frac{\sin \pi n}{\sin \pi x} \right| \le \frac{1}{|\sin \pi x|} = O\left(\frac{1}{\|x\|} \right),$$

$\|x\|$ indicating the distance to the nearest integer. □

Bibliography

[Ap1] T. M. Apostol, Remark on the Hurwitz zeta function, *Proc. Amer. Math. Soc.* **2** (1951), 690–693.

[Ap2] T. M. Apostol, On the Lerch zeta-function, *Pacific J. Math.* **1** (1951), 161–167.

[Ap3] T. M. Apostol, Addendum to 'On the Lerch zeta-function', *Pacific J. Math.* **2** (1952), 10.

[Ap4] T. M. Apostol, *Introduction to analytic number theory*, Springer, 1976.

[Ber6] B. C. Berndt, Identities involving the coefficients of a class of Dirichlet series VI, *Trans. Amer. Math. Soc.* **160** (1971), 157–167.

[Ber3] B. C. Berndt, On the Hurwitz zeta-function, *Rocky Mount. J. Math.* **2** (1972), 151–157.

[Ber4] B. C. Berndt, Two new proofs of Lerch's functional equation, *Proc. Amer. Math. Soc.* **32** (1972), 403–408.

[Ber2] B. C. Berndt, The gamma function and the Hurwitz zeta-function, *Amer. Math. Monthly* **92** (1985), 126–130.

[Böh] P. E. Böhmer, *Differenzengleichungen und Bestimmte Integrale*, Koecher Verlag, Berlin, 1939.

[Bor] J. M. Borwein and P. B. Borwein, *Pi and the AGM: A study in analytic number theory and computational complexity*, Wiley, 1987.

[BG] P. Bateman and E. Grosswald, On the Epstein zeta-function, *Acta Arith.* **9** (1964), 365–373.

[BKT] R. Balasubramanian, S. Kanemitsu and H. Tsukada, Contributions to the theory of Lerch zeta-functions, to appear.

[BKY] R. Balasubramanian, S. Kanemitsu and M. Yoshimoto, Euler products, Farey series, and the Riemann hypothesis II, *Publ. Math. (Debrecen)* **69** (2006), 1–16.

[Ca] R. Campbel, *Les intégrales Eulériennes et leurs applications. Étude approfondie de la fonction gamma*, Dunod, Paris 1966.

[Car] L. Carlitz, A note on the Dedekind sums, *Duke Math. J.* **21** (1954), 399–403.

[Com] L. Comtet, *Advanced Combinatorics: The Art of Finite and Infinite Expansions*, Reidel, Dordrecht, Holland 1974.

[CS] S. Chowla and A. Selberg, On Epstein's zeta-function (I), *Proc. Nat. Acad.*

Sci. USA **35** (1949), 371–374; *Collected Papers of Atle Selberg I*, Springer Verlag, 1989, 367–370. *The Collected Papers of Sarvadaman Chowla II*, CRM, 1999, 719–722.

[Da] H, Davenport, *Multiplicative number theory*, Markahm 1997, second edition Springer, 1982.

[D] C. Deninger, On the analogue of the formula of Chowla and Selberg for real quadratic fields, *J. Reine Angew. Math.* **351** (1984), 171–191.

[EM1] O. Espinosa and V. H. Moll, *On some integrals involving the Hurwitz zeta-function: Part 1, The Ramanujan J.* **6** (2002), 159–188.

[EM2] O. Espinosa and V. H. Moll, *On some integrals involving the Hurwitz zeta-function: Part 2, The Ramanujan J.* **6** (2002), 449–468.

[Erd] A. Erdélyi, W. Magnus, F. Oberhettinger and F. G. Tricomi (The Bateman Manuscript Project), *Higher Transcendental Functions*, Vol. **I**, McGraw-Hill, New York, Toronto, and London, 1953.

[Fine] N. J. Fine, Note on the Hurwitz zeta-function, *Proc. Amer. Math. Soc.* **2** (1951), 361–364.

[Fu] T. Funakura, On Kronecker's limit formula for Dirichlet series with periodic coefficients, *Acta Arith.* **55** (1990), 59–73.

[GR] I. S. Gradshteyn and I. M. Ryzhik, *Table of integral series and products*, Academic Press, New York etc. 1965.

[GZ] M. L. Glasser and I. J. Zucker, Lattice sums, *Theoretical Chemistry: Advances and Perspectives*, Vol. 5, ed. by D. Henderson, Academic Press 1980, 67–139.

[Hata] M. Hata, *Real analysis in number theory — Problems and Solutions*, to appear

[Hashimoto] M. Hashimoto, Examples of the Hurwitz transform, to appear.

[HKT] M. Hashimoto, S. Kanemitsu and M. Toda, On Gauss' formula for ψ and finite expressions for the L-series at 1, to appear.

[Hautot] A. Hautot, A new method for the evaluation of slowly convergent series, *J. Math. Phys.* **15** (1974), 1722–1727.

[H] A. Hurwitz, Einige Eigenschaften der Dirichlet'schen Funktionen $F(s) = \sum \left(\frac{D}{n}\right) \cdot \frac{1}{n^s}$, die bei der Bestimmung der Classenanzahlen binärer quadratischer Formen auftreten, *Zeitschrift f. Mathematik u. Physik* **27** (1882), 86–101.

[Is] M. Ishibashi, An elementary proof of the generalized Eisenstein formula, *Sitzungsber. Österreich. Wiss. Wien, Math.-naturwiss. Kl.* **197** (1988), 443-447

[IK] M. Ishibashi and S. Kanemitsu, Dirichlet series with periodic coefficients, *Res. Math.* **35** (1999), 70–88.

[KTTY4] S. Kanemitsu, Y. Tanigawa, H. Tsukada and M. Yoshimoto: Some aspects of the modular relation. Number Theory: tradition and modernization, ed. by W. Zhang and Y. Tanigawa, (Developments in Mathematics, Vol. 15) Springer, (Febuary 2006) 103-118

[Ka] S. Kanemitsu, *On evaluation of certain limits in closed form, Théorie des Nombres*, J.-M. De Koninck and C. Levesque (éds.), 1989, 459–474, Walterde Gruyter 1989.

[KKaY] S. Kanemitsu, M. Katsurada and M. Yoshimoto, On the Hurwitz-Lerch zeta-function, *Aequationes Math.* **59** (2000), 1–19.

[KKSY] Kanemitsu, S., Kumagai,, H., Srivastava, H. M. and Yoshimoto, M., Some integral and asymptotic formulas associated with the Hurwitz zeta-function. *Appl. Math. Comput.*

[KTTY1] Kanemitsu, S., Tanigawa, Y., Tsukada, H. and Yoshimoto, M., On Bessel series expressions for some lattice sums: II. *J. Phys. A: Math. Gen.* **37** (2004) 719–734.

[KTTY2] Kanemitsu, S., Tanigawa, Y., Tsukada, H. and Yoshimoto, M., Crystal Symmetry Viewed as Zeta Symmetry. Proceedings of Kinki University Symposium "Zeta Functions, Topology and Quantum Physics 2003" (Developments in Mathematics, Vol. 14) Springer (April 2005) 91–129

[KTTY3] Kanemitsu, S., Tanigawa, Y., Tsukada, H. and Yoshimoto, M., Contributions to the theory of the Hurwitz zeta-function. to appear.

[KTY7] S. Kanemitsu, Y. Tanigawa and M. Yoshimoto, Ramanujan's formula and modular forms, in *Number-theoretic methods — future trends* (ed. by Shigeru Kanemitsu and Chaohua Jia), Kluwer Academic Publ., 2002, pp. 159–212.

[KTY1] S. Kanemitsu, Y. Tanigawa and M. Yoshimoto, Structural elucidation of the mean square of the Hurwitz zeta-function, *J. Number Theory* **120** (2006), 101–119.

[KTY2] S. Kanemitsu, Y. Tanigawa and M. Yoshimoto, Determination of some lattice sum limits, *J. Math. Anal. Appl.* **294** (2004), 7–14.

[KTZ] S. Kanemitsu, Y. Tanigawa and W.-P. Zhang, On Bessel series expressions for some lattice sums, to appear.

[KTZ2] S. Kanemitsu, Y. Tanigawa and J.-H. Zhang, Evaluation of the Spannen-integrals of the product of two zeta-functions, to appear.

[Kan] S. Kanemitsu, Some sums involving Farey fractions, RIMS Kôkyûroku **958** (1996), 14–22.

[KKY1] S. Kanemitsu, M. Katsurada and M. Yoshimoto, On the Hurwitz-Lerch zeta-function, *Aequationes Math.* **59** (2000), 1–19.

[KKY3] S. Kanemitsu, H. Kumagai and M. Yoshimoto, Sums involving the Hurwitz zeta function, *Ramanujan J.* **5** (2001), 5–19.

[KKY2] S. Kanemitsu, H. Kumagai and M. Yoshimoto, On rapidly convergent series expressions for zeta- and *L*-values, and log sine integrals, *Ramanujan J.* **5** (2001), 91–104.

[Kat] M. Katsurada, An application of Mellin-Barnes' type integrals to mean square of Lerch zeta-functions, *Collect. Math.* **48** (1997), 137–153.

[Kat1] M. Katsurada, Power series and asymptotic series associated with the Lerch zeta-function, *Proc. Japan Acad. Ser. A Math. Sci.* **74** (1998), 167–170.

[KM] M. Katsurada and K. Matsumoto, Explicit formulas and asymptotic expansions for certain mean square of Hurwitz zeta functions I, *Math. Scand.* **78** (1996), 161–177.

[Klu] D. Klusch, On the Taylor expansion of the Lerch zeta-function, *J. Math. Anal. Appl.* **170** (1992), 513–523.

[Ko] N, S. Koshlyakov Investigation of some questions of analytic theory of the rational and quadratic fields, I-III (Russian), *Izv. Akad. Nauk SSSR, Ser. Mat.* **18** (1954), 113–144, 213–260, 307–326; Errata **19** (1955), 271.

[LG] A. Laurinchikas and R. Garunkstis, *The Lerch zeta-function*, Kluwer Academic Publ., Dordrecht-Boston-London 2002.

[Lan] E.Landau, Vorlesungen uber Zahlentheorie Bd II, Leipzig1927=Chesea ??

[Leb] N. N. Lebedev, Special functions and their applications, Dover 1972.

[Le] M. Lerch, Note sur la function $K(w, x, s) = \sum \frac{e^{2\pi k i w}}{(w+k)^s}$ *Acta Math.* **11** (1887), 19–24

[Leh] D. H. Lehmer, Euler constants for arithmetic progressions, *Acta Arith.* **27** (1975), 125–142; Selected Papers of D. H. Lehmer, Vol. **II**, 591–608, Charles Babbage Res. Center, Manitoba, 1981.

[Leh2] D. H. Lehmer, A new approach to Bernoulli polynomials, *Amer. Math. Monthly* **95** (1988), 905–911 =Selected papers of D. H. Lehmer,

[LDH] H .-L. Li, L.-P. Ding and M. Hashimoto, *Structural elucidation of Eisenstein's formula*, to appear.

[Li] R. Lipschitz, Untersuchungen der Eigenschaften einter Gattung von unendlichen Reihen, *J. Reine Angew. Math.* **105** (1889), 127–156.

[LT] H.-L. Li and M. Toda, Elaboration of some results of Srivastava and Choi, *J. Anal. Appl.* **25** (2006), 517–533.

[Ma] C. J. Malmstén, De integralibus quibusdam definitis, seriebusque infinitis, *J. Reine Angew. Math.* **38** (1849), 1–39.

[Matsumoto] K. Matsumoto, Recent developments in the mean square theory of the Riemann zeta and other zeta-functions, Number Theory ed. by R. P. Bambah et al, Hindustan Books Agency, 2000, 241–286.

[Me] Hj. Mellin, Die Dirichletschen Reihen, die zahlentheoretischen Funktionen und die unendlichen Produkte von endlichem Geschlecht, *Acta Soc. Fenn.* **31** (1902), 1–48.

[M1] M. Mikolás, Mellinsche Transformation und Orthogonalität bei $\zeta(s, u)$; Verallgemeinerung der Riemannschen Funkutionalgleichung von $\zeta(s)$, *Acta Sci. Math. (Szeged)* **17** (1956), 143–164.

[M2] M. Mikolás, Integral formulae of arithmetical characteristics relating to the zeta-function of Hurwitz, *Publ. Math. (Debrecen)* **5** (1957-58), 44–53.

[M3] M. Mikolás, A simple proof of the functional equation for the Riemann zeta-function and a formula of Hurwitz, *Acta Sci. Math. (Szeged)* **18** (1957), 261–263.

[M4] M. Mikolás, New proof and extension of the functional equation for Lerch's zeta-function, *Ann. Univ. Sci. Budapest* **14** (1971), 111–116.

[Mi] J. Milnor, On polylogarithms, Hurwitz zeta functions, and the Kubert identities, *Enseign. Math. (2)* **29** (1983), 281–322.

[Ni] Nielsen, N., *Traité Élémentaire des Nombres de Bernoulli*. Paries: Gauthier-Villar 1923.

[Ob] F. Oberhettinger, Note on the Lerch zeta function, *Pacific J. Math.* **6** (1956), 117–120.

[Pa] A. Papoulis, The Fourier integral and its applications, McGraw-Hill, 1962.

[PK] R. B. Paris and D. Kaminski, *Asymptotics and Mellin-Barnes Integrals*, Cambridge University Press, 2001.

[PP] Cheng-Dong Pan and Cheng-Biao Pan, *Elements of Analytic Number Theory*, Science Press, Beijing, 1991.

[R] H. Rademacher, *Topics in Analytic Number Theory*, Springer-Verlag, Berlin, 1973

[Sa] F. Sato, Searching for the origin of the theory of PHV (Prehomogeneous Vector Spaces), *Annual Meeting of the Math. Soc. Japan* 1992.

[Sch] O. Schlömilch, Uebungsaufgaben für Schuler, Lehrsatz von dem Herrn Prof. Dr. SCHLÖMILCH, *Arch. Math. u. Phys. (Grunert's Archiv)* **12** (1849), 415.

[Sla] L. J. Slater, *Confluent Hypergeometric Functions*, Cambridge University Press, Cambridge, London, and New York, 1960.

[SC] A. Selberg and S.Chowla, On Epstein's zeta-function, *J. Reine Angew, Math.* **227** (1967), 86–110; *Collected Papers of Atle Selberg I*, Springer Verlag, 1989, 521–545; *The Collected Papers of Sarvadaman Chowla II*, CRM, 1999, 1101–1125.

[Su1] Z.-W. Sun, On covering equivalence, in "Analytic Number Theory," Kluwe Academic Publishers 2002, 277–302.

[Su2] Z.-W. Sun, Curious identities and congruences involving Bernoulli polynomials, to appear

[Ter1] A. Terras, *Harmonic Analysis on Symmetric Spaces and Applications I, II*, Springer Verlag, New York-Berlin-Heidelberg, 1985.

[Tit] E. C. Titchmarsh, *The Theory of the Riemann Zeta-Function*, Oxford University Press, 1951; revised version by R. D. Heath-Brown, Oxford University Press, 1986.

[UN] K. Ueno and M. Nishizawa, *Quantum groups and zeta-functions*, in *Quantum Groups: Formalism and Applications*, Proc. of the Thirtieth Karpacz Winter School (Karpacz, 1994) (J. Lukierski et al., Editors), pp. 115–126, Polish Sci. Publ. PWN, Warsaw, 1995.

[Wa] K. Wang, Exponential sums of Lerch's zeta functions, *Proc. Amer. Math. Soc.* **95** (1985), 11–15.

[Wal] A. Walfisz, *Gitterpunkte in Mehrdimensionalen Kugeln*, Polish Sci. Publ. PWN, Warsaw, 1957.

[We] A. Weil, On Eisenstein's copy of the Disquisitiones, *Advanced Studies in Pure Mathematics* **17**, 1989 Algebraic Number Theory – in honor of K. Iwasawa, pp. 463–469

[Wil1] J. R. Wilton, A proof of Burnside's formula for $\log \Gamma(x + 1)$ and certain allied properties of Riemann's ζ-function, *Messenger Math.* **52** (1922/1923), 90–93.

[Ya] Y. Yamamoto, *Dirichlet series with periodic coefficients*, Proc. Intern. Sympos. "Algebraic Number Theory", Kyoto 1976, 275-289. JSPS, Tokyo 1977.

Index